I0814447

BLACK GIRLS GARDENING

BLACK GARDE

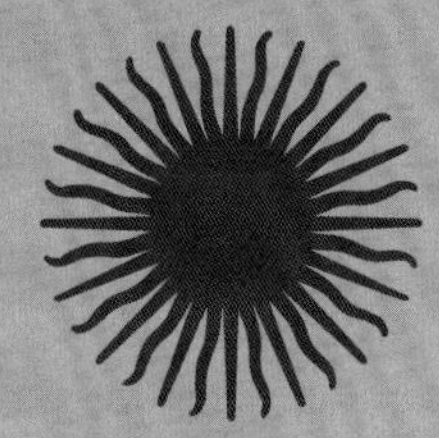

EMPOWERING
STORIES and
GARDEN WISDOM

for Healing and
Flourishing in Nature

AMBER GROSSMAN

GIRLS
ENING

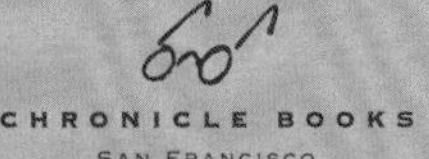

Library of Congress Cataloging-in-Publication Data available.

ISBN 978-1-7972-2824-2

Manufactured in China.

Design by Rachel Harrell.
Typesetting by Wynne Au-Yeung.

10 9 8 7 6 5 4 3 2 1

Chronicle Books LLC
680 Second Street
San Francisco, California 94107
www.chroniclebooks.com

This book is dedicated to my grandmother, Anne Stafford, a kind and caring human. Thank you for always being my inspiration to be a better person than I was yesterday. I appreciate your constant love and support with everything I do. May this make you proud.

CONTENTS

HOW IT ALL BEGAN

For as long as I can remember, all I ever wanted to do was be outside. Some of my earliest and favorite memories begin with foraging for wild blackberries in the bushes on our property line, running (unapproved) laps through Grandpa's tall corn stalks in his backyard garden, and playing barefoot in the yard with my brother on warm summer days until the sun went down. Nature has always felt like my playground and a place I'm connected to.

I grew up in a small town in upstate New York where the population was under ten thousand residents. The first place I remember living was in an older two-story white house that sat on top of a hill about twenty minutes outside of town. It had a long, rocky driveway lined by apple trees. To the right of the driveway, just before you'd get to our house, sat a large red barn where my grandpa stored his tractors and other equipment. I lived there only as a child, but I think of it often. In typical upstate New York fashion, our rural property had a large open field in the backyard that met the tree line. Across the driveway was a small pond plentiful with fish. This setting provided different landscapes to explore, but one common theme was apparent among them: They were full of life. I observed bees, butterflies, and birds flitting about, minding their work. Although we didn't have much of a garden there, it became clear to me that I wanted to cultivate and connect with nature.

While my parents were off to work, my brother and I spent most of our summers and free time at our grandparents' house. They had a stand of fruit trees that always seemed to exist in abundance. My brother and I would get so excited to climb them to pick fresh apples and peaches. Grandma would gather the peaches we had picked, and we would sit with her in the kitchen, eager to help her in the canning process. The fruit trees stood guardian over their impressive garden, which my grandfather gave a lot of attention to. He would call us over to the picnic table to share watermelon slices and laughs with him. Witnessing my grandfather's love of tending to his garden, the peace it evoked, the effect it had on his state of mind and mood now feels like the catalyst for my love of getting my own hands dirty. I may not have realized it then, but these moments were truly the beginning of the lifestyle I've built for myself today. Growing up in the country gave me a deep appreciation for the inexpressible feeling of being outdoors surrounded by nature, but it was the time spent at my grandparents' that made me want a yard full of fruit trees and a vast garden I could pick from whenever I wanted.

As the years went by, I decided it was time to leave New York. I found myself in a few different cities before settling down in Wilmington, North Carolina. My husband and I purchased our first house in the fall

cucumber

of 2017, and less than six months later, I was ready to get started on my first garden. Though I didn't have much knowledge of how to actually build or grow anything, I knew I needed to just go for it. Starting is the hardest part! I found some old 2 by 4 boards and spare nails lying around the shed and thought *Hmm, this could make for a perfect raised garden bed!* After choosing some of my favorite vegetables, I stuck a couple of seeds in the soil and hoped for the best. In a few short weeks, I already had cucumbers, peppers, and tomatoes sprouting, along with a few different herbs. After two months, I was able to have a small weekly harvest. It was highly rewarding. I found so much joy in watching the plants I had started from seed grow and thrive. Having access to fresh produce encouraged me to learn more skills in the kitchen so that my hard work would not go to waste. I've learned how to complete the cycle of gardening, harvesting, and enjoying the fruits of one's labor, and then returning the inedible parts back to the land as compost.

As I got into my second year of gardening, I became laden with issues that I had no idea how to handle. Even though I was growing the same vegetables and herbs that had thrived the year before, I started to deal with pest issues and plant diseases I knew nothing about. That's when I began my search for some inspiration online. It wasn't long before I discovered the large gardening community that exists on social media, but even more obvious to me was the lack of Black women being represented on a lot of popular platforms.

In the summer of 2019, after continuously looking for more women to connect with, I decided to create the platform BlackGirlsGardening on Instagram. The sole intention was to network with other Black female gardeners. I envisioned that the account could be a space where Black women could have a sense of common ground to connect and influence one another to get into gardening. After a few short months, I noticed how many Black female gardeners and farmers there already were, with years of experience under their belts. They had skills and pearls of wisdom to share, so I slowly transitioned the account to not only share inspiration for new gardeners but also showcase those who have already been established for several years or even decades. Over time, and through our shared love for gardening and the outdoors, I formed deeper connections and friendships with some of these women. It showed me that even though we are usually privately tending to our gardens, there is a whole community out there that is doing the same and looking for others to share their experience with.

For me, growing up in a small town as a minority made finding similar women to relate to that much more important. I wanted to make sure I created a space that was inclusive to the women I felt were being left out. My original plan was to try to find about five thousand Black women and share their passion for growing food and tending to plants via social media. The more I posted these women, the more followers I gained, and they would message me about how excited they were to finally find other similar women to connect with. This really gave the account a purpose. The more I heard stories from all over, the more it inspired me to want to try new things in my own garden, like planting varieties of vegetables and fruits that I had never heard of before.

This book compiles short stories I've gathered by Black and Mixed-Black women on how they began their journeys with gardening, what their experiences have been like, and what gardening means to them. They share personal details and insightful tips intended to help you reconnect with your birthright—the powerful autonomy of connecting with the land. There are chapters on growing food, gardening with kids, and backyard, flower, and community gardens, so there's something here for all types of gardeners! Each chapter also includes advice on different aspects of gardening, such as composting and companion planting. My hope is that this book will inspire you to plant a few seeds wherever you are and see how easy it can be to turn your dreams of growing your own garden into a reality.

DIVERSITY
PATIENCE
PEACEFUL
COMMUNITY
JOY

CHAPTER 1

BACK
GAR

YARD

DENS

A PROMISE OF BOUNTY

ASHLIE THOMAS
@the.mocha.gardener

Ashlie Thomas is the owner of The Mocha Gardener and author of *How to Become a Gardener: Find Empowerment in Creating Your Own Food Security.*

I have always had an interest in nature and health and wellness, and in many ways I've carried this passion with me throughout my personal and professional life. Therefore, prior to the conception of our garden, I certainly had a defined reason for growing food, and this absolutely fueled what would become my purpose and mission, working toward greater food security, especially in underserved communities. I love what gardening advocate Lauri Krantz says: "When you feel like everything is out of control, what nurtures better than planting a seed with the promise of bounty and the health that comes with it?"[1] I held this sentiment when I began my garden years ago, and it still resonates just as deeply for me today.

So how did this all begin? In 2018, my husband and I began envisioning what was possible for our yard. We wanted to create not only a space for growing healthier food options for ourselves and our family but also a grounding space for relaxing and reflection—simply put, a green safe haven. For months, we walked around the backyard of the one acre around our mid-century bungalow in North Carolina, examining the environment, watching the movement of the sun, and mentally designing the layout of our future garden. During this planning period, there were several logistical aspects we needed to consider before getting started. Aspects such as infrastructure, materials, environmental factors, and even our own capacity at the time. We worked through all of these things diligently and over time. Now, I must admit that we went big with our first year, which officially began in the spring of 2019. My husband, with his creativity and building skills, took about five days to design and

1. Jeanette Marantos, "When the World Feels Scary, I Want to Garden. Here's What to Plant Right Now," *Los Angeles Times*, March 11, 2020, https://www.latimes.com/lifestyle/story/2020-03-11/what-to-plant-in-southern-california-right-now

build a total of nine wood-sided garden beds, while I constructed a game plan for what we would grow and how we were going to be successful. Thankfully, our first gardening season *was* a success, and even more important, we learned so much about ourselves as gardeners and how we work together as partners.

Since we started our garden in 2019, much has evolved in the space. Over time, we've added structures like a she-shed for processing garden produce, a greenhouse for seed starting, a custom garden arbor for the entrance, in-ground garden beds, and lastly a chicken coop for our egg-laying chickens. We've also grown a variety of vegetables, fruits, trees, bushes, flowers, and culinary and medicinal herbs over the years, and we've done so year-round. So I like to say that our garden has become more seasoned, and ultimately so have we as growers. For several years we have cultivated and nurtured this sanctuary, and in return, it has cultivated and nurtured us—even in the midst of some of the most historically chaotic times in our world.

My gardening journey has been remarkable in that I (along with my family and community around me) have been blessed with so many unimaginable tangible and intangible fruits of this labor. And for this, I feel immense gratitude each day.

LIVING OFF-GRID

SHAQUITA JACKSON
@thegreendesert

Shaquita Jackson is a nature lover, mom of two, wife, producer, storyteller, influencer, and educator. Her passion for sustainability drove her to cofound GreenDesert.org, a platform that for over eighteen years has been inspiring people to live a self-sufficient and green lifestyle.

Imagine a small hobby that has the power to change your life. That's exactly what happened to me more than twenty years ago when I discovered gardening. What started as a simple pastime turned into a lifestyle that has transformed the way I see the world and the way I live my life. It has given me a sense of empowerment and helped me connect with the natural world in a way that is both therapeutic and magical. Over time, I have learned the importance of using sustainable practices, and how to reduce my carbon footprint by mimicking Mother Nature.

My family and I built and transformed two shipping containers into off-grid homes and then transferred them from Arizona to Texas. In Arizona, we aim for zero waste; while in Texas, where we spend several months each year, we are entirely off-grid. We harvest our own seeds. We make our own fertilizer and nutrient-rich soil using things like our chickens' eggs for calcium, their poop for nitrogen, and compost from our kitchen scraps, yard waste, and even paper products. We try our best to avoid throwing away anything that can harm the environment. We collect rainwater and use some of it to water our garden; the rest we clean with sand filters for drinking. The benefits of all of this led to our becoming more self-sufficient, helping fortify us against economic downturns and inflation. By growing our own food, we have a steady supply of fresh, flavorful, and organic produce that I know is free of harmful pesticides and chemicals. My garden has become my personal grocery store and pharmacy, saving me thousands of dollars per year, and because there are no "food miles" (that is, the distance food travels from where it's grown to your kitchen) involved, the food is always at its peak freshness. I also benefit from growing medicinal herbs, like moringa, chamomile, and hibiscus, which I use to make teas and natural remedies for common ailments.

As someone deeply connected to the earth, I find that gardening is a powerful way to nourish not just my body, but also my mind and soul. It's helped me heal physically and mentally, providing free therapy, meditation, and education. Through this lifestyle, I have connected with others, shared our harvest with family and friends, helped children plant gardens at their schools, and spread love with our GMO-free seeds. The garden has become a sanctuary that brings my family peace, joy, and a sense of freedom. It's a place where I go to disconnect from the stresses of everyday life and focus on the things I can control, like nurturing a plant from seed to harvest.

Gardening has truly been a journey filled with rewards. Whether it's the taste of fresh eggs from your own chickens or the feeling of empowerment that comes from growing your own food, gardening has the power to change your life.

GREENHOUSE DREAMS

URSULA CARMONA

@CarmonaAcres

@HomeMadebyCarmona

Ursula Carmona is the creative mind behind the home and garden blog *Home Made by Carmona*. When she isn't gardening, she is sharing creative DIY tutorials and helpful tips for decorating your home. You can find her gardening hacks on social media at @CarmonaAcres, and her DIY and design tips at @HomeMadebyCarmona.

I've always believed that working the soil and spending time in nature was wonderfully beneficial, but I didn't truly fall in love with gardening until I slowly started turning my yard into my own private oasis. Who knew adding twinkly lights and planting flowers in between the vegetables could turn a garden into an experience rather than a chore!?

I distinctly remember the moment I caught the gardening bug. One morning, I was gazing out my bedroom window at the garden area, and said to my husband, "I've never built a structure, and I don't know how, but in a couple of weeks, I'll be walking into a greenhouse I designed and built with my own two hands!" I'd always imagined having a greenhouse but didn't have the budget to purchase one new. Luckily, soon after that I came across some twenty-year-old windows slowly rotting away in a neighbor's yard, just waiting to be unearthed and turned into that greenhouse I'd dreamed of. Two weeks and two bruised hands later, I was beginning to fill my new greenhouse with plants and already dreaming up the next garden project. I began to treat my yard as an outdoor room, slowly designing the space with as much care as I would devote to my interiors.

While the greenhouse may have sparked my romanticism toward gardening, horticulture/backyard planting certainly isn't new to me. When I was growing up in the high desert of Southern California, my parents turned our backyard into an oasis you could spot long before you reached the house. Ours was the house that had a forest crammed into the three-quarter-acre backyard, complete with fruit trees, grapevines, and climbing trees, in addition to a vegetable garden. A true wonder in a place with actual cacti and tumbleweeds that looked like it was out of a Western. Looking back, I've truly come to appreciate how much work they put into making the yard a magical space for me and my siblings. I hope one day my kids also will

look back and treasure the memories of our time spent in the garden.

My approach to gardening has changed over the years, and I've learned to adapt during different seasons of life. When my children were little, I felt like a garden failure because I couldn't seem to create what I envisioned the perfect garden should look like: even, clean rows of vegetable plants. It took me years to begin to realize there are a myriad of gardening methods and styles, and I could do what worked best for me. Now my garden tends to be more of a wild beauty, with raised beds that are packed to the brim, and I love every bit of it! It's not unusual to see vines trailing above, creeping herbs spilling over the border of my beds, and loads of flowers right there among my vegetables. But the secret sauce to my garden are the perennial flowers and herbs that anchor the garden beds. They keep things beautiful and green when I can't devote as many hours as I would like to gardening. Perennials come back year after year on their own, and I interplant seasonal vegetables all around them.

I would like to see more people turn gardening into an experience. One that is more than just planting seeds and weeding. Cultivate a charming space with cafe lights, trellises, and a place to sit and gaze at the fruit of your labors. If you turn your garden into an enchanting space, then you'll spend as much time enjoying it as you do laboring in it.

CORONA

RECLAIMING MY RELATIONSHIP WITH THE LAND

LEOTA WILSON

@curlycultivators

Leota Wilson is a queer, Black gardener who grows fruit, vegetables, herbs, and flowers. In the past several years, she and her partner have transformed their barren backyard into an urban food forest. Leota's mission is to inspire her community to grow some of their own food.

I grew up gardening alongside my mother. I loved being right at her elbow, soaking in all of her knowledge and enjoying her abundant harvests. As I got older, gardening was just a pleasurable hobby for me. However, today it has turned into so much more. It is a form of therapy, a lesson in patience, and a way to sustain myself. It is also a way that I get to connect with my partner, Mikaela.

As the daughter of a South Carolinian sharecropper, I understand that gardening is about more than just growing food and flowers. For me, it's about honoring my ancestors, reclaiming my relationship with the land, and finding joy. I am committed to minimizing my ecological footprint and adopting more sustainable practices in my everyday life. As I did research and became more informed about the impact of pesticides and herbicides on our environment and bodies, I dedicated myself to using organic practices in my garden. I am passionate about food justice, and I love educating the community on how to grow organic produce in their own space.

In the past few years, I have completely transformed my California backyard into an urban food forest. With the privilege of growing food year-round in Sacramento, I rarely need to purchase vegetables or fruit from the market. I do believe there is a deep connection between food and culture, so I enjoy growing culturally relevant crops, such as okra, fish peppers, and collard greens; but I also grow a variety of other vegetables along with stone fruits, berries, citrus, herbs, and a variety of flowers.

At one point during the COVID-19 pandemic, I realized that I poured so much into my garden; I am very mindful of how plants respond to stressors and their environment, but I noticed I was less mindful of that in myself. Because of my garden, I began to reframe the way I care for

myself—checking in with myself, practicing mindfulness, and pouring into *myself* with intention. This experience reminded me of the power of self-care. My garden is a true reflection of who I am. I consider myself a lifelong learner, and gardening allows me to be curious and explore.

The garden has become a place where we make our dreams a reality, through research, goal setting, and hard work. I would not be the person I am today without this lifestyle and all the lessons it has taught me. It humbles me to know that there will never be a time when I can consider myself an expert gardener, since there is always something to learn.

PLANNING OUT A GARDENING SITE

Whether you're a novice gardener, a professional landscaper, or someone in between, there are a few fundamental things to consider when planning a new gardening site or expanding a current one. One of the most important is having an idea of the plants you'd like to grow. (See page 175 to learn more about USDA zones.) Once you know that, you'll understand what kind of requirements your garden needs to meet, like the amount of sunlight and water, and the type of soil the plants will grow best in.

- **LIGHT:** The majority of plants that we put in our gardens desire at least six to eight hours of direct sunlight a day. The vegetables and fruits that typically require this amount of light produce fruits from the flower of the plant (think tomatoes, cucumbers, strawberries, squash, raspberries, and so on). There are, of course, several vegetables that can also be grown in partially shaded areas, like arugula, lettuce, spinach, broccoli, cauliflower, and cabbage, to name a few. By first deciding which plants you would like to grow, you can better plan where in your garden they'll get the light they need. Observe the area in the morning, mid-morning, late afternoon, and all the way until the sun begins to set. Make a chart of where the sun is during each of these hours; is the spot you're considering for your garden getting full sun, partial shade, or full shade? It is a best practice to observe the potential garden space during each season, as the angle of the sun changes throughout the year. Observation is the first step on the path toward abundance! Observing your land in this way will help you feel more connected with and responsible for your space.

- **SOIL:** Every gardener knows that soil quality is key to having a healthy, happy garden. If the soil quality is poor, your harvests will likely be disappointing. You may find your plant leaves beginning to turn brown or yellow, which is an indicator that you may have an issue. The typical array of vegetables and fruits that we put in our gardens love well-draining, fertile soils. Well-draining soil means that water can soak into the roots, then drain to a level beneath them; in dense clay soils it will drain too slowly, and in sandy soils it will drain too quickly. If you are unsure of your soil quality, you can do a soil test at home with a kit or send a sample to a local lab for testing. This usually costs only a few dollars or may even be free. If the soil quality test results show poor soils in your yard, consider using raised beds, containers, or grow bags. You can mix together your own organic material and compost to fill them with a nutrient-rich base for your plants to grow in. Another option is a no-dig garden: Instead of digging into the dirt, you create the beds or rows by layering organic materials—like cardboard, dead leaves, and plant debris—along with a layer of compost directly on top of the current soil. Though you can plant directly into the compost with this method, I've personally found my harvests to be better when I mix the compost in with the soil as well.

- **WATER:** Two important aspects of easy gardening are placing your garden where you can readily access it, and having a watering source nearby. Work smarter, not harder! Typically, the best time to water your plants is early morning to early afternoon. This gives the plants and soil plenty of time to dry before the sun goes down. This is desirable because when water sits at the roots or on plant foliage overnight, it encourages fungal growth and a place for insects to lay their eggs. Keeping the garden close to where you live or work allows you to check in frequently for issues so you can start to treat any that arise before they become a serious problem.

Choosing your garden site can be the most daunting step of the whole process, but try not to overthink it. Remember that this is all just a fun science experiment and an experience meant to enrich your life. Simply placing a few pots of herbs in a sunny windowsill can foster a closer connection with plants and lighten the heart. However, if you do want to go big and plant a garden, approach it with curiosity. The role of the steward, though serious, cultivates a playful and fun relationship with plants. Armed with an attitude of wonder, there is no failure—only lessons. Your lessons will carry over from year to year as you get to know your garden through close observation and care. You just have to decide where to plant it!

A LINEAGE OF LANDOWNERS AND GARDENERS

CHRISHA FAVORS
@naturally_chrisha

Chrisha Favors is an educator, land steward, gardener, forager, self-proclaimed naturalist, and homesteader of her thirteen-acre property in Eugene, Oregon. She has been an advocate for nature since a young age and hopes to inspire future generations to cultivate the same joy she experiences in nature.

When I was a young girl, I was ripe with curiosity and had a steadfast drive to connect with nature. I grew up in southern Georgia, where I discovered the joys of gardening and stewarding land with my grandfather as my personal mentor. His Indigenous knowledge and insight inspired me to continue his legacy of being a gardener, loving land steward, and friend of nature. Some fond childhood memories that continuously come up for me involve my time on his property, frolicking through his garden, shucking peas for dinner, chewing on raw sugar cane straight from the garden, foraging for blackberries, and eating fresh garden-to-table mustard greens.

Gardening is a birthright for me. I come from a generation of landowners and gardeners, and it is an honor for me to continue passing down the knowledge my Black and Indigenous ancestors gave me. I consider gardening a reclamation of my ancestral wisdom; it is a pathway for me to relate to my ancestors in a wholesome way. I love how this simple act of living connects me to something bigger than me.

I've experienced some tragic moments in my life and have been able to utilize the garden as a place of solace. It has taught me many lessons throughout my journey: Patience, a sense of reciprocity, and a nurturing approach are key. I also find that my capacity to love living beings in this world has expanded since gardening in my adulthood. Cultivating plants from seeds or starts requires an immeasurable amount of wait time until they bear fruit. That type of patience creates a sense of caring, of guardianship, and a respect for the process. The feeling it spawns in me is honestly very personal and maternal. I believe reciprocity with nature is essential to creating a respectful and long-lasting relationship with the land, and I am grateful for the mentorship I received from my grandfather to continue cultivating that relationship.

My love for gardening also extends to my nieces and nephew, who love gardening with me. They even have their own little plots at their homes, and it brings me such joy. I feel honored to be able to pass down my knowledge to the next generation in my family line and to continue the cycle of healing. As I grow older, I recognize my role as a future ancestor. Gardening is a tradition in my family, and I plan to continue this practice and inspire others to get their hands dirty and connect with the land. I truly find joy, healing, and connection in the garden, and I hope that my presence in this space uplifts generations to come.

NURTURING MY BODY, NURTURING MY SOUL

HANNAH VEGA
@afroforagers

Hannah Vega is a passionate gardener, forager, and nature photographer based in Maryland. With a deep connection to nature, she's cultivated a diverse garden and honed her skills in wild plant identification, gathering an array of delicious, nutritious, and medicinal treasures from the land.

Stepping into the vibrant tapestry of my wild garden, I can't help but marvel at the beauty of nature and the bounty it provides. As a Black woman deeply rooted in nature's embrace, my journey through the green world has been one of self-discovery and empowerment, and a tribute to the resilience and self-sufficiency of my ancestors.

I've been innately curious about and drawn to the natural world for as long as I can remember. Growing up, I didn't have a garden. But I was regaled with stories about some of my ancestors who worked the land, had gardens, could identify all the plants in the woods, and knew how to use them. These stories fascinated me, gave me a sense of pride, and became a guide for the kind of knowledge that I wanted to acquire. When I married and moved out of my childhood home, I started my first miniature container garden, using discounted, wilted plants from my local organic market that I brought back to life. I instantly fell in love and knew I had found my calling!

There was (and still is) a lot of trial and error in gardening, but I loved how much I was learning about this new hobby. While growing into my new garden life, I also took an interest in learning about the wild edible and medicinal plants growing freely in my environment, which shattered my misconceptions about "weeds." I discovered that so many plants society calls "weeds" are nutritious, delicious, and/or medicinal!

Fast-forward to my life now. It's evident that gardening and foraging have become much more than hobbies for me—it's a lifestyle. My garden is not the manicured and predictable landscape often found in suburban neighborhoods, but an ever-changing wild sanctuary where nature reigns. There's a combination of hardy culinary and medicinal perennial plants as well as annual flowers, vegetables, and herbs. A few of my favorite plants are elderberries, heirloom Appalachian "greasy beans," hot peppers, watermelons, passionflower, comfrey, and cherries. I have also kept space for many "weeds" I eat or use medicinally, such as dandelion, yarrow, stinging nettle, lamb's quarters, chickweed, and plantain leaf.

In a world where industrialization and technology have distanced us from our natural origins, and the wisdom of the land is often disregarded, my work as a gardener and forager has become a journey to reclaim traditional knowledge that has been obscured or forgotten over time. My wild garden and foraging practices have enriched my life beyond measure. They have taught me patience, mindfulness, resilience, and the beauty of simplicity. But most importantly, they have given me a deeper understanding of my heritage and the knowledge that, as a Black woman, I am intrinsically connected to the earth and all its gifts. Through my garden, I am not only nurturing my body but also nourishing my soul.

THE CINDER-BLOCK GARDEN

ASIA SPRATLEY
@yellowdoorurbanhomestead

Asia Spratley is an urban homesteader with a focus on gardening. She works full time and is a mother of two.

While there is something totally romantic about the slow country life, there's also still something that keeps me in my urban setting. What piques my interest related to gardening and animals has little to do with location. My connection to food and nature arose right where I was, in my small urban backyard in Virginia. My vintage home is situated on an average-sized urban lot. If you were to visit, you'd walk into a cozy garden full of cattle panel trellises, vegetable plants, herb bushes, and fruit trees. The garden occupies more than half of my backyard. As your gaze moved across the garden, you'd notice a small corner dedicated to the feathered residents. My urban homestead is currently home to seven chickens and five quail.

In my adult life, I've found that I need to have a hobby or project going. In 2020, when the world was forced to stay home, I needed something to do that kept me home but not inside. I had passively grown food before as a fun activity with my daughter. In mid-summer, after watching a few YouTube garden videos, I decided to relocate a small cinder-block bed from a shady spot into a sunnier section of the yard. This relocation turned out to be the start of a beautiful hobby-turned-way-of-life.

During the off-season, I used YouTube, the internet, and books to grow my knowledge. I couldn't wait for spring to make its way back around to me. I anticipated the fullness of the cinder-block raised beds, the beautiful harvests of heirloom tomatoes, eating fresh green beans, and watching my daughter enjoy a fresh, homegrown cantaloupe. I was ready to plant hope and watch it grow.

Today that one relocated cinder-block garden bed still remains in my garden. The dimensions have changed a bit. It serves as a reminder of how I started gardening, and it continues to grow our food. I'm appreciative of those few cinder blocks and the

WELCOME

soil that changed the way I feed my family. When I think of all the things my garden provides, food isn't always at the top of my list, although it's definitely the obvious answer. To be honest, the garden itself, the physical location, gives me a place to be busy while also reflecting. You'd be surprised how many conclusions I've come to in the garden and how many times it has changed a bad mood into a grateful one. My son often refers to my garden as my "happy place," and he is right! The process of gardening allows me to have wins that I am proud to share with the world. It allows me to feed my family food that I know was grown with love, naturally. It stretches me outside of my comfort zone to be just a little bit better than last year, a little more confident than the year before, and a little more grateful than ever. It's been said that all experts were once beginners. Eventually, I'll be someone's definition of an expert, while continuously learning. There's always something new to learn in a garden.

BURPEE

HOMESTEADING IN OHIO

CIEARRA EVANS
@thethriftedplanter

Ciearra Evans is a gardener and homesteader in Ohio. She has been gardening for over ten years.

Plants have always been a part of my life, even when I didn't realize it. I grew up in a house full of plants my mother carefully tended to. Every summer, she would fill her garden with flowers. As I got older, I would receive houseplants as gifts and would promptly kill them and not think twice about it. I even had a vegetable garden that I let get so overgrown and unruly I stopped tending to it. It wasn't until I had a home of my own that I gradually started taking an interest in growing my own plants. I quickly filled my home with gorgeous foliage; soon after that, my obsession spilled into landscape plants and then vegetables. I began binge-watching gardeners on YouTube and then found the plant community on social media.

After several years, my partner and I wanted to dive further into homesteading. We didn't have space for large animals but still wanted to do something more. Our solution was chickens. Chickens provide so much for our home and garden. They make great little composters, eating up all of our leftovers and garden discards. At the end of the season, the chickens get all of our uneaten vegetable plants. I pull everything right out of the ground and toss it into their run. Like little machines, they devour and break down the plant matter into manure and eventually compost. After several months, we go in and "harvest" the composted material to use in our garden. Our chickens also reward us with delicious and healthy eggs and entertain us with their silly antics.

Since I garden in a colder gardening zone, it's important for me to start my vegetable and herb seeds early and indoors. This process can start as early as January for certain plants that I want in my garden. At this stage of my gardening obsession, I start 90 percent of my own vegetables and herbs from seed. I find seed starting to be so exciting! I love seeing those first signs of growth; they give me a boost of adrenaline to keep on going when winter feels endless.

As I have grown in the garden and homestead, my needs and focus have changed. I have changed. I need my garden to grow organic and medicinal vegetables and herbs. I need my garden to grow beautiful flowers and foliage. Growing my own vegetables and garden plants from seed has been a lifestyle change for me. Focusing on healthy foods for my family has become a priority. What I love most about gardening is that there is always something to learn and discover. What worked one year may not work the next. The garden connects us to nature and to each other. It's a place of peace, wonder, inspiration, and most of all, growth.

GARDENING TOOLS THAT MAKE LIFE EASIER

Using our hands to pull weeds and cultivate the soil is therapeutic; it allows us to form a deeper, more personal connection with the land. But as our gardens expand, we also want to be smart about how we use our time and energy. There are an unimaginable number of tools we can use to make our lives as gardeners much easier and our chores less time consuming.

Sometimes we try to garden on a budget (which I am a huge fan of), but over the years I've learned that quality is the most important thing. Gardening doesn't need to be an expensive hobby or lifestyle, but if your equipment is well made and durable, you won't have to be constantly replacing your tools and wasting money. Invest in good tools, and then search local thrift stores for garden goodies like storage containers, pots, decor, and harvest baskets.

The following are some tools that I've found to be super helpful when doing daily gardening chores.

- **GARDENING HOSE/DRIP IRRIGATION:** Watering can be time consuming, especially if you have a lot of plants. In a smaller space like a patio or small container garden, a two-gallon watering can will work just fine. When you're working in a larger space, you will want a good hose and you might even consider a drip irrigation system (if that's in your budget). Drip irrigation systems are small hoses connected to your water spigot and placed in pots or laid out throughout your garden beds/rows. They release drips of water directly at the roots, reducing evaporation and conserving water. They can save you a lot of time spent watering, as you can either set it up with a timer or just turn it on and go about your other tasks!

- **GARDEN TROWEL:** A trowel is a small hand-sized shovel. It is one of my favorite tools, as it makes it easy to break up the ground and dig the perfect-size hole for planting seeds or transplants. There are several different styles of trowels; I prefer ones with a rubber handle grip over a wooden handle, as they are much more comfortable and save your hands in the long run! Another useful version comes from Japan: The hori hori is a blade-edged trowel, great for digging in hard ground.

- **HAND SHEARS:** This tool allows you to prune new growth or dead leaves from plants with ease and precision. The most common types of hand shears are anvil and bypass; the big difference is their blade styles. Anvil pruners come in handy when cutting woody stems or dead branches, while bypass pruners are better suited for young growth or thinner stems. You can even use hand shears to experiment with pruning your plants into fun shapes.

- **WHEELBARROW:** Depending on the size of your garden, a wheelbarrow can be your best friend. They can haul hundreds of pounds in a single load—making it easier to move compost, soil, mulch, or any garden debris you may need.

- **GLOVES:** Typically I'm digging in the dirt with my bare hands, but I have found that wearing a good pair of gloves for more laborious tasks like garden cleanup or pruning helps protect your skin from cuts, blisters, calluses, and insect bites. To give your hands protection from thorns or sharp branches, wear gloves made of a thicker material like leather. However, always consider forgoing gloves as often as possible to receive the health benefits of touching soil. Did you know that there are tiny bacteria in the soil that tell your brain to release serotonin through contact with your skin? The soil wants a relationship with you, and it needs your bare skin to form one!

CHAPTER 2

GARD
KITC

EN TO

HEN

TWO LEGACIES

ERIKA DUDLEY
@erika_francoitalian

Erika Dudley is a historian, artist, community organizer, and chef who is forever curious. With her mom, Claudette, she is cofounder of Onion Dip for Breakfast, a media project highlighting food, gardening, the sea, art, travel, and family. Her academic focus is French history and Italian; she was educated at Harvard University (French history) and Le Cordon Bleu Paris.

Several years ago, I created the Black Garden, a physical green space that centers plants and produce that are associated with Black people or black in color, and a figurative space for exploration and rest that honors our ancestors. It reflects two legacies: traveling and gardening. Nearly every plant in my fertile patch is inspired by the tiny and grand gardens I've seen around the world.

My mom's side of the family has been traveling to every continent except Antarctica for five generations with the express desire of exploring the beautiful, curious, and delicious. Visiting Japanese and Italian gardens led me to plant a pomegranate bush and jet-black eggplants here in Chicago. Red and black currants recall small cottages and castles in France. Green stems shooting up in my beds take me back to China and North Africa. Tender, juicy fraises des bois (wild French strawberries) now grow all over my garden and remind me of bending down and foraging with my toddler in the Swiss Alps. We are also a family that pickles garden vegetables like okra and onions and makes jars of the sweetest garden fruit jams. My own preserves from the garden are the fondest homage to and remembrance of family. If I make a pastry filled with these preserves, my grandmama's hand is in it.

My father's side of the family nurtured fig tree clippings from the very plantation where our ancestors were enslaved. My grandmother, a child of farmers, regaled me with family stories of 1920s rural Georgia through her life in bustling Atlanta. She couldn't resist taking a clipping from any plant that had a striking flower, beautiful leaf, or unusual shape. She would add it to her lovely, sprawling garden of roses, hydrangeas, pineapples, gardenias, and lilies. She especially loved cake, and my "garden cakes"—tender desserts with layers of homemade garden liqueurs and purées topped with flowers and herbs—serve as mementos crisscrossing through time. The

common threads are familiar. Liberated Black bodies and minds travel the world, grow plants for beauty, nurture life from seed to fruit, pass on expertise, look to the future, explore and experiment just because, and inspire possibilities. My parents modeled this beautifully.

My gardening encourages abundance: a bouquet of herbs and collards, a botanical cocktail in a carafe, a tart laden with tomatoes, or a confection covered with fruit. I have been growing and making my whole life, and every time I harvest from the garden and cook or bake, I tighten my link to my mom, grandmama, grandmother, and all the greats.

This pavlova represents the best of the Black Garden. It is light and airy, brightened by sweet berries, tart currants, earthy flowers, homemade inky liqueurs, and aromatic herbs. It is grounded in family customs and techniques. It somehow takes me back to foreign lands while being deeply rooted in Black traditions.

WHERE FOOD FOR THE SOUL IS CULTIVATED

NAIMAH ABRAHAM
@nimzyunmeasured

Naimah Abraham is a Haitian American mother of two with over thirty years of experience gardening. She is a self-taught baker and home cook who focuses on farm-to-table meals that fuse multicultural techniques and cuisines. She is also a best-selling author and teaches seasonal cooking and gardening workshops.

When I was a very young girl, my mother introduced me to gardening after we moved to our family home on Long Island. She taught me the ins and outs of gardening: how to use a trowel to dig a hole the proper depth, how to place and space your plants, when and how to water, and the basics of landscape design. As the years passed, I grew up alongside the garden. Every spring I'd go to the nursery with my mother and select seedlings for our kitchen garden. In the summer, I would create harvest boxes of the bumper crops to give away to family, friends, and neighbors. In the fall, I would collect and dry herbs for preservation. Preceding winter's arrival, I'd dig up and salvage any plants before the first frost—something I learned from my paternal grandmother, who would walk with me and tell me in her native Haitian Creole which plants needed to be brought indoors.

In 2010, when I became a mother, gardening took on an entirely deeper meaning for me. My son had food allergies and sensitivities. I used my garden harvests to make his baby food myself. This allowed me to play a more active role in his nutrition and to better control what he was exposed to. Through motherhood, my garden transformed from an accessory to a necessity. Meals for my family became more vibrant and elaborate because I had fresh crops on hand. Having a garden challenged me to learn about different applications of plants and to grow more food—which in turn encouraged my culinary curiosity and new gastronomical escapades. Now I could benefit from the entire plant instead of using just the fruit or vegetable. For example, I harvest figs from my tree to eat fresh or make jam. I also use fig sap as rennet to make ricotta and soft cheeses. The fig leaves can be used for tea or as a garnish to beautify a dish.

Gardening has allowed me to connect to some aspects of my heritage as a Haitian American that would otherwise be lost.

I can grow herbs and crops that have cultural significance for my family but are not readily available in US markets. I grow borlotti, or cranberry beans, which are very hard to find fresh in American markets, to make *diri ak pwa* (Haitian rice and beans). When I gave birth, my grandmother and great aunt prepared tea and herbal baths using herbs from my garden that are traditionally used in Haiti to promote postpartum healing and breast milk production. To be able to preserve this rich ancestral knowledge is a privilege that gardening has afforded me.

Over thirty years of gardening later, I'm still growing and creating new and delicious meals. Every season I grow something experimental, a novel crop that I've never grown before. When I prepare a farm-to-table meal for my family using my own harvest, I am not just Mom but also a farmer, chef, artist, and creator. I find that deeply fulfilling. This garden of mine contains a treasure box of vivid core memories filled with seeds of love. My garden is where family, faith, and food for the soul are cultivated.

CANNING AND PRESERVING

BRITTNEY HUTCHERSON
@themelaninurbangardener

Brittney Hutcherson was born and raised in Milwaukee, Wisconsin. She started her garden back in 2014 and has been actively gardening ever since.

When I was about five or six, my dad and I would go every Mother's Day weekend to a local nursery and buy flowers to plant in our front yard. We would purchase anywhere from ten to fifteen flower flats to fill up all of our planters and the in-ground area in front of our house. According to my dad, I would help him for only the first two to three hours, and after that I would disappear. (I don't remember it going that way, but he's probably right.) I always remember us buying marigolds and petunias, and to this day, I purchase them every year for my garden.

As I grew up, I always loved cooking, but making my own pickles, infused olive oils, and herb salts has taken my culinary skills to the next level. Buying food from the grocery store is okay, but the taste of a fresh heirloom tomato, your own crushed red peppers, or homemade fresh giardiniera is unmatched. The process of canning and preserving can be time consuming, but it's always worth it in the end. The only reason I grow cucumbers is because I enjoy pickles and making my own. I make about five types of pickles a year: regular dill, old-fashioned dill (I leave the herbs and spices in the jar), sweet heat, bread and butter, and extreme heat dill pickles (made with Carolina Reaper peppers). I also make four different infused olive oils from the herbs and veggies in the garden; the garlic-infused olive oil is my absolute favorite.

The vegetable I grow the most of in my garden is tomatoes. On average, I grow 140 tomato plants a year. About 90 percent are heirloom varieties; the other 10 percent are your standard paste tomatoes, like Roma. My favorite category of heirloom tomatoes are blue/black beauty varieties. Something about the tanginess of the darker parts of the tomatoes made me fall in love at first bite. I use them in everything from fresh pico de gallo to sauces to fire-roasted tomatoes.

My current garden is a mix of mainly vegetables and herbs that I enjoy using in my dishes or preserving. There is also a mix of annual and perennial flowers for my pollinator friends. Plants come in so many shapes, sizes, and colors that you can design a garden that looks any way you can imagine. Through cultivating my garden, growing my own food and bringing it to the table, I've been able to create the lifestyle I want.

COMPOSTING FOR BEGINNERS

A great way to start shrinking our carbon footprint and reducing the amount of waste being sent to the landfill is composting. Simply defined, composting is the gathering and storing of organic matter so it can decay and be used to return nutrients to the soil. It's a natural cycle that improves the health of your soil, your produce, and, in turn, your body.

With any method you choose, you'll want to make sure you have a mix of brown and green ingredients. Brown ingredients are carbon-based materials like dead leaves, twigs, hay, straw, shredded newspaper, plain brown cardboard boxes, and even household paper scraps like paper towels or toilet paper rolls. Carbon is necessary for the compost to work—it's the basic building block in creating new microbes and the energy source for the microbes to eat as they break down the materials. Green ingredients include fruit and vegetable scraps, potato skins, eggshells, coffee grounds, tea (removed from the bag), grass clippings, old flowers, and plant cuttings. The green ingredients are high in nitrogen and provide most of the nutrients to the compost. A common starting ratio is two parts brown ingredients to one part green ingredients, but you can play with this.

There are certain things you should always avoid putting in your compost piles or bins, like meat or cooked foods, grease, dairy products, invasive weeds, highly acidic fruits, or pet waste. The smell of some of these things can bring unwanted visitors to your garden, and they can also contaminate your compost pile if they contain any harmful bacteria.

Here are a few different common methods for composting.

- **COMPOST TUMBLER:** Some home gardeners prefer to use a composting tumbler set outside in the sun. The sun's heat helps break down the organic material faster, and tumbling the receptacle further speeds up the process. Also, it's fun and satisfying to spin the tumbler!

You can also use any type of large plastic container or storage bin to save on cost. Just drill a few holes on all sides for air circulation, fill it with a mix of your materials, pop the lid on, and make sure to give it a stir every few weeks.

- **COMPOST PILE:** Another method often used at home, which requires almost no tending, is creating a large pile by just putting all your composting materials in one area on the ground. You'll want a nice sunny spot for this as well, since your kitchen scraps and the carbon you put in need sun heat to fully decay, break down, and rebuild in time for planting. This method, also called cold composting or passive composting, does tend to take a bit longer than using a tumbler, but it will still give you the same end result.

- **CITY-SPONSORED COMPOSTING PROGRAMS:** Some cities have programs that allow you to pay for a rented compost bin. Once you fill it with all your kitchen scraps and materials that can be composted, they will pick it up and leave a clean bin for you. Although this is a great way to dispose of compost material, you won't actually receive any compost to use in your garden.

- **COMMUNITY GARDEN:** You can either participate in a community garden (these usually have compost piles) by renting a garden plot, or drop off your compost materials there (with permission). This is a good option for those who don't have a garden at home but still want to their compostable materials to be used for gardening rather than throwing them into the garbage.

Composting completes the cycle of gardening. It is the decay part of the cycle—which restarts the whole process. How magical it is to watch your kitchen scraps transform into rich black soil that you can spread around your garden and use to enhance the life of your plants!

NATURE'S ORGANIC KITCHEN

KOLA STARNES
@luvya_garden

Kola Starnes is a certified garden consultant who opened Luvya Garden in 2021. She helps transform urban yards into beautiful organic vegetable kitchen gardens! She also specializes in consulting and offers coaching, planting plans, and workshops.

The sunny days and high temperatures here in Lancaster, California, make growing a vegetable garden a little tricky, but I'd always dreamt of having a backyard garden. I wanted to grow enough of my favorite organic vegetables to eliminate purchasing produce from the grocery store. After a few years of trial and error, I was ready to give up, until one day I finally got it right! Once I got the hang of growing food in Lancaster's unique climate, my garden began to thrive. I now produce more food than my family can consume, so I share it with loved ones. It makes me feel so good to grow wholesome organic produce and feed my family.

My current organic kitchen garden is an extension of our home. We have seven 7 by 4 foot, 2 foot high custom raised beds with a curated organic soil blend. I wanted to grow as much food as I could in our raised beds, so I decided to grow vertically to save space. Cherry tomatoes and cucumbers grow up two beautiful metal arch trellises and two obelisk trellises. Flowers grow at the end caps and borders of the beds to bring in the pollinators, deter insects, and add a pop of color and beauty. I intensively plant a variety of seasonal vegetable plants throughout the raised beds. Gravel pathways suppress weeds and give a clean look. I incorporate sustainable practices in my garden by making my own fertilizer by composting and vermicomposting vegetable scraps, eggshells, coffee grounds, and yard waste.

When I harvest an abundance of vegetables, I like to preserve them long term by freezing, canning, pickling, or drying them. Preserving food captures fresh produce at the peak of flavor and enables me to prepare a homemade meal with homegrown ingredients for my family, rather than running to the grocery store or eating out. Canning tomato sauce for marinara allows us to experience the taste of garden-fresh tomatoes all year round, and canning jars stay good in the pantry for two years.

I blanch and freeze collards, mustards, kale, zucchini, and green beans; these last ten to twelve months. Quick-pickled cucumbers will keep, refrigerated, up to three months. I preserve herbs by tying small bunches with string and hanging them to dry.

I also decided to raise chickens for eggs. Our initial flock of four chickens quickly grew to eight, and now I collect about eight dozen eggs monthly!

I think of my kitchen garden as an oasis. A magical place of peace and harmony. It's uplifting, relaxing, and guaranteed to put a smile on my face. It feeds my soul with gratitude and respect for Mother Nature.

My advice? Start small, have patience, and enjoy the process. Plant vegetables that you enjoy eating and that grow well in your area. Know what to plant when. Feed the soil with nutrient-rich compost and worm castings, and the soil will feed your plants. Always remember, don't compare your day one to someone else's year ten.

IN THE KITCHEN, PREPARING GARDEN-FRESH MEALS

BATAVIA CHOCTAW
@b_bettagarden

Batavia Choctaw is a client relationship executive residing in Chicago, Illinois. She fell in love with gardening over fifteen years ago after carving out a little space in her backyard to grow a few vegetables.

In 2007, during my house-hunting process, my real estate agent and I visited a quaint little house about three miles from where I grew up. Before I walked into the house, we went to the backyard, where there was a 30 by 4 foot strip of mud covered in weeds and early fall leaves. I looked at my agent and said, "I could plant a garden here." This was a surprise to him, because a backyard, let alone a garden, was not on the must-have list I'd given him for my search. I too was surprised, because I had only two memories of gardening, which I would not describe as fun-filled! You see, there was the six-year old Batavia who had tightly clutched her grandmother's hand as she looked around for garter snakes while her grandfather worked in the community garden plot. And then there was the eight-year-old Batavia who was not allowed to go outside to play with friends until she had finished shelling peas and snapping beans.

The summer after purchasing that house with the muddy patch of dirt and leaves, I announced to my family that I would be planting a garden! My grandparents came over to my new home to help me plant tomatoes, cucumbers, and peppers. My grandfather showing me how to work up the soil and my grandmother insisting that our cucumber mounds were not high enough—these remain some of my favorite garden memories. That summer of 2008 would not be the last time my grandparents and I would connect through the garden. Years later, my grandfather helped me build my first raised bed. He would visit and plant collard greens; I would go back to my family home to help my grandmother plant flowers (my first love). And many years later, we would be connected again as I leaned on the garden to help me through the grief that followed the passing of each of my grandparents.

It has been more than fifteen years since that first summer garden, and not only has my garden footprint grown, but I too have grown as a gardener. Most seasons you

can find me growing vegetables, herbs, and a few fruit plants in more than sixteen raised beds and thirty containers across my backyard and front yard. Each year, my small garden space in the middle of a big city produces hundreds of pounds of food that I can enjoy throughout the year by way of dehydrating, fermenting, freezing, and canning. In the early years, I would eat most of my garden harvest during lunch or dinner throughout the week, and I would freeze the extras. While I still enjoy my harvest in weekly meals during the early part of the growing season, my garden-to-kitchen experience has changed significantly as the garden season progresses. The busiest and most rewarding time of year for me is when summer turns to fall and I can split my time between outdoor garden activities and preservation projects in the kitchen. When the garden starts to really produce late in the summer, you will find me in the kitchen daily preparing garden-fresh meals and preserving pepper jelly, salsa, hot sauce, collard greens, potatoes, green beans, tomato sauce with homegrown garlic and herbs, and more! Growing healthy, nutritious food and finding a way to enjoy it beyond the harvest is a life skill. Continuing to hone this skill and helping others become the gardener they may not know they can be, is where I find my garden joy!

HOMEGROWN COLLARDS FOR THE HOLIDAYS

ALNISSA RUIZ-CRAIG
@alnissa.grows

Alnissa Ruiz-Craig is a wife and mother of three. She is an organic gardener in suburban Georgia, zone 8a, with a passion for growing her own food and flowers.

While I didn't have any family members who had a vegetable garden, my grandmother had an incredible green thumb when it came to plants and flowering shrubs. As I started to witness a shortage of fresh produce in the stores during the pandemic (including my favorite herb, cilantro), I decided to see if I'd inherited that green thumb by giving vegetable gardening a try. I started out with a couple of basic gardening supplies, a handful of knowledge (from YouTube), and an abundance of enthusiasm. That enthusiasm soon transformed into a true passion and prompted me to develop a deeper connection with the earth and the food I eat.

My garden serves as a source of pride when I can pull ingredients right from my own backyard. As someone who loves to cook (and eat), I grow all kinds of fruits and vegetables and enjoy growing things that I wouldn't typically find at the local grocery store, like heirloom tomatoes, purple sprouting broccoli, kohlrabi, and Kajari melons, to name a few. I also love experimenting with foods I've never had. Before I started growing my own, I hadn't even heard of Swiss chard; now it's one of my favorite greens and a staple in my garden. If you haven't tried an omelet with sautéed chard, garlic, fresh herbs, and a little goat cheese, I highly recommend it!

While I love using ingredients from my garden for my everyday cooking, using them for family gatherings and holiday dinners has been extra rewarding. Thanksgiving and Christmas are both special times of year when my passion for gardening truly pays off. My first fall/winter garden was full of different varieties of brassicas, specifically collard greens—from 'Old Timey Blue' to 'Champion' and of course Georgia's own 'Georgia' collards. It was an amazing feeling to know how I nurtured those tiny seeds from late summer, and come Thanksgiving and Christmas day, they were the stars of the show on the holiday dinner table. The smiles on my family's faces were a direct reflection of the love I pour into my garden.

Cooking and preparing the fruits of my labor is the most gratifying part of the gardening process, but sometimes my harvests cannot be consumed right away. As a gardener, you are forever a student, learning new skills, techniques, and best practices. I am working on improving my food preservation skills to make the most of my harvests. Currently, I enjoy pickling and freezing many of my vegetables for later use, and dehydrating herbs to create my own spice blends.

My very first garden consisted of five containers. It has since exploded into what I consider to be my own little personal oasis. I went out on a limb and dedicated half of my small yard to my garden. With an elaborate layout that consists of twelve raised beds, trellises, arches, two tower gardens, and multiple decorative containers, it is a place that I am proud of and brings me joy. As I continue to learn from my successes and failures and share my gardening journey, it is my hope to inspire and spark an interest in gardening for others.

GREENHOUSE GARDENING FOR BEGINNERS

Having a greenhouse is not essential to starting or having a garden. However, it can extend a gardener's growing season by several months, and it makes for a dreamy addition to any area. A greenhouse allows certain plants to grow year-round in regions where they typically wouldn't be able to, as it provides a climate-controlled space perfect for starting seedlings and growing plants that need protection from certain elements like frost, wind, and predators.

Depending on the size of the greenhouse you build or purchase, it can accommodate an array of plants in trays or pots. If your greenhouse doesn't have a floor, you can even plant directly into the ground to utilize all the space. Bigger is not always better; sometimes it's better to start small and figure out the ins and outs before taking on something at a larger scale! Whatever you use, be sure it's sturdy and can hold up to the weather in your area. Living on the coast of North Carolina with the chance of hurricanes, I chose to go with a medium-sized, steel-framed greenhouse with the base cemented to the ground.

The greenhouse should be placed in a sunny area of your yard or patio, as your plants will need the warmth of the sun to survive. Ventilation is key, especially during the summer months when the trapped heat can become unbearable and there's the risk of frying your plants. I recommend using a thermometer to keep track of the temperature inside the greenhouse. In warmer months, open any doors or windows, or have fans running to keep the air circulating. In the winter, you can place a small gas/propane heater inside (never blowing directly onto your plants) to keep the temperatures up above freezing. I also discovered that bubble wrap works really well for insulation in the winter; just affix it to the walls with waterproof tape, and lay it on the ground beneath your plants. I never buy bubble wrap, but I do save sheets from old packages and use those!

Some gardeners turn their greenhouse into a personal sanctuary, filled with not only plants but also decor and joy-inducing trinkets. Pick a style that works for you and your space, and make it a cozy place to relax while soaking in the view of your hard work. You want it to be an environment that you actually want to spend time in. It'll be a nice escape from the chaos of the day, where you can find true "me" time to relax, unwind, and connect with the plant kingdom.

You may be wondering *Well, what can I even grow in a greenhouse?* The short answer is, pretty much anything! I have grown so many different plants in my 6 by 7 foot greenhouse, from cauliflower, potatoes, onions, tomatoes, and peppers to an extensive variety of herbs and flowers. I always say, when you want to grow something but you don't know where to start, begin with the things that you like to eat or you enjoy looking at! What are some vegetables or herbs that you regularly buy at the store? Or perhaps you want access to certain produce that you cannot buy in your area. Got an old family recipe calling for a plant not found at your market? With a greenhouse, you can grow these yourself. Growing plants you are personally connected to is a great way to develop a bond with your garden. As long as you research the basic needs of the plants—how much lighting they require, how much space they need, and the temperature required for them to germinate—you can confidently grow just about any plant life in a greenhouse.

WHEN TO PLANT WHAT IN YOUR GREENHOUSE

- **EARLY SPRING:** Cold-hardy crops like lettuce, spinach, carrots, beets, and radishes.
- **LATE SPRING:** Brassicas like cauliflower, broccoli, and cabbages. Start seedlings for heat-loving, long-season crops such as tomatoes, peppers, beans, melons, or eggplant.
- **SUMMER:** If your greenhouse is big enough and not too hot, you can still grow your heat-loving crops (tomatoes, peppers, beans, melons, eggplant). However, ventilation and humidity are important factors to ensure they don't wilt or overheat.
- **FALL:** Herbs and veggies like arugula, Swiss chard, kale, garlic, and Brussels sprouts. Bring into the greenhouse any potted plants or citrus trees that may be endangered by frost.
- **WINTER:** Usually a time for dormancy in most gardens, but you'll still be able to harvest plenty of the leafy greens and herbs you planted in the fall. Plan to eat lots of winter salads!

CHAPTER 3

FLO

GAR

WER
DENS

WARD'S FARM

ALYSSA WARD

@lyssacouture & @wardsfarmnj

Alyssa Ward is an operations manager at a compounding pharmacy from 9 to 5 and a flower farmer the rest of the day. She lives in South Jersey with her husband, Allen, and dog, Junior.

I have always had a deep fascination with the growth and development of plants. I grew up in the Poconos and have fond memories of assisting my mother in her garden during the warmer months. As I helped her with watering, she would always advise me to count to ten for each plant pot. During my college years, I grew sunflower seeds in my dorm room. I was a biology major, and my favorite course was botany, where we had the opportunity to create our own unique plant. Even as a young adult, I had a pineapple plant in my apartment.

When I met my husband, Allen, in 2013, he was growing organic vegetables on a small-scale farm called Ward's Farm NJ. To add color and happiness to the field, he would line it with sunflowers. Dahlias are my favorite flower, so we decided to start growing them as well. At that time, we lived in a populated area in South Jersey and set up a roadside stand in front of our house where we would set out our extra veggies and flowers. We noticed that many days our veggies would remain but neighbors scooped up all the flowers.

In 2018, we moved to an eleven-acre farm. That is when we converted from growing organic vegetables to cut flowers. Now we focus on selling wholesale to florists, creating custom bouquets, offering flower shares, and providing bulk flowers for DIY brides and party planners. We grow a variety of cut flowers including gladiolus, rudbeckia, echinacea, cosmos, and zinnias, with a special focus on field-grown flowers. As it's an organic flower farm, we prioritize sustainability by implementing practices such as planting diverse cover crops to replenish soil nutrients, prevent erosion, foster beneficial insects that control pests, and suppress weed growth. We are proud to avoid the use of chemical pesticides and fertilizers, and we do not use plastic wrapping for our bouquets.

In 2019, we opened our farm to the public with "pick your own" events. Our main goal was to share the farm experience with our community. These events had no entry fee, and visitors paid for only what they picked. Throughout the years, we have hosted various events, such as Yoga on the Farm in collaboration with a local yoga studio, and beer and cheese tastings with a local beer tour company. Recently, we have introduced new and exciting events like BYOB Date Night and Picnic Day, which have quickly become our favorites. These events allow visitors to enjoy the farm's peaceful and serene atmosphere.

Our farm has become a special place for people to create unforgettable memories. With photo shoots from first dates to engagements and anniversaries, newborns to first birthday cake smash celebrations, we have been a part of it all. It is an honor to know that our farm is a part of these special moments for our visitors.

djolus $1 Per Stem
Cash$
In the

DINING ALFRESCO

JULIA BENN
@julbeartgardens

With a strong background in art, design, and gardening, **Julia Benn** is a former art gallery and design studio director who marries traditional and modern to create beautiful interiors and outdoor living spaces that stand the test of time.

Growing up on a farm in the early '70s in Barbados, watching my late grandfather, an agriculturalist and master gardener, had a profound and lasting impression on me. My grandmother, also a lover of plants, had a lady of the night (night-blooming jasmine) planted near her home's entrance; the aroma was pleasantly intoxicating. She grew the most stunning, enormous coleus plants and colorful Joseph's coat. As I plant coleus in my gardens, and any time I smell jasmine, I think warm thoughts of her. Gardening is deeply ingrained in my DNA, for my ancestors on both my mother's and father's sides were avid gardeners, so as I garden, I give thanks and honor them.

Some of my favorite childhood memories were harvesting Barbados cherries, guavas, golden apples, mangoes, pawpaws, soursop, sugar apples, tamarinds, and ackee. I also recall picking wildflowers, particularly West Indian lantana and marigold, which we referred to as "Stinking Missy." Today, I incorporate marigolds among my flower and veggie beds, as they attract beneficial insects and repel the harmful ones. Additionally, their intense scent assists in keeping the inquisitive visiting deer at bay, as woodlands surround my Maryland garden.

Aside from flowers' natural appeal and stunning beauty, I cultivate them to attract native bees and pollinators. They also let me connect with my neighbors as I share my harvest. I love many flowers, but hydrangeas and peonies are two of my favorites.

I integrate plants and flowers into all my design projects, whether interior design, event decor, or outdoor living spaces. They enhance the color scheme, hold sentimental value, and are infused into the most extraordinary, memorable moments of our lives.

Dining alfresco is another favored pastime, so the outdoor dining room is a significant feature in my garden (setting a gracious table is a skill I learned from my late mother). I enjoy cooking and hosting dinner parties, especially using fresh produce and flora from the garden. Considering scale, texture, and space planning, I integrate my design skills to create captivating, comfortable, and inviting outdoor living spaces. In addition to dining, there are also zones for reading and relaxing in my garden.

Much more than merely beautifying the landscape and growing produce, my flower and vegetable gardens provide solace, peace, and calm, a safe reprieve from the outside world. And gardening keeps me physically fit and grounds me both literally and figuratively. Gardening and being at one with nature gives me tremendous joy, and I am eternally grateful for inheriting this gardening gene and gift!

FLOWERS BELONG EVERYWHERE

JANICE GROVES
@jaysgardenjournal

Janice Groves is a self-taught gardener who loves to create garden beds by sheet mulching. She grows a variety of flowers and vegetables in South Carolina, zone 8.

When I was growing up, my parents always spent time in the garden, but I had no real interest in it. What I did love was eating the delicious tomatoes that my father grew each summer, and watching my mother's eyes light up as she examined the new growth on her camellia bushes.

I took up gardening when I moved into my own home in 2005. My lot is one-fourth of an acre situated in a cookie-cutter subdivision in South Carolina. Gardening here, in zone 8, allows for a long growing season. Originally my garden consisted of an over-treated lawn, two invasive Bradford pear trees, and boxwoods along the foundation. This was very typical landscaping in the early 2000s, but I knew that I wanted my yard to eventually be more productive. I wanted to harvest vegetables and beautiful flowers within steps of my home.

Part of my planning process, then and now, consists of staring at the yard and imagining what it could look like in the future. I knew that I needed to take my time and let the garden evolve. I frequently walked through the local nurseries, learning about different flowers that I could add to my garden, and each weekend I looked for new clearance plants to fill in certain spaces.

I began to slowly extend my existing garden beds after every rainfall. (Since my land is mostly heavy red clay soil, waiting for rainfall made it easier to dig out the edges of the garden bed.) I made note of the existing walkways and of the natural paths that I usually took through the garden to determine where to add new beds. I also created a long, deep garden bed along the perimeter of the yard. This allowed me to have a natural fence around my property, providing some privacy. I made sure to include at least one tree and several shrubs to serve as the "bones" of the garden bed. Then I added lots of perennials and flowers.

I love having an array of colors in my garden, but for each bed, I try to limit the

color palette slightly. My front garden has a bed of hot colors: orange, red, and yellow. The side yard features mostly green and burgundy foliage with pops of white from my panicle hydrangeas, while my island bed consists of plants with really large leaves such as canna lilies, a rice-paper tree, banana trees, and hardy ginger.

Some of my favorite flowers to grow are actually herbs that I allow to flower, such as sage, basil, mustard greens, spinach, and rosemary. I think these flowers are so beautiful and serve double duty in the garden as edible and ornamental plantings.

In addition to providing a creative space for myself, the garden has become my healing space. When my mother died, I found myself comforted by the sounds of the birds singing in my garden. To this day, each morning, with coffee in hand, I take time to do a walking meditation, while always looking for new growth and examining the health of my plants. I spend this time daydreaming about new projects I want to start in the garden. I have a small budget but big dreams. I want to encourage others to garden with what they have right now. Don't wait! Be willing to make mistakes and learn from them. Enjoy the garden in every season, and use the lessons from the garden to appreciate the changes in daily life. Happy gardening!

COMPANION PLANTING

Companion planting is the practice of growing two or more plant types near each other so either or both receive some type of benefit. An approach as simple as growing nectar-rich flowers in between plants to help attract pollinators to nearby fruit and vegetable plants is a great example.

This technique comes to us from many different Indigenous peoples of North and Central America. In their Three Sisters companion planting of corn, beans, and squash, each plant performs an essential function for the others: the beans fix the nitrogen in the soil to enhance the growth of the corn, whose stalks provide a structure for the beans to climb; the large leaves of the squash create cool, moist soil for all. Together they create a sustainable and fertile ecosystem in the soil, as well as a healthy diet that Native Americans have subsisted on for thousands of years. Thanks to their ingenuity, and similar wisdom from farming cultures around the world, we can all benefit from companion planting.

The better we understand our garden and plant needs, the easier it is for us to make clear choices on where to place them. But it can often be hard to tell which plants will pair well together based on looks alone. Ask your fellow gardeners and do a little research. Every season I learn something new, and I always like messing around with different plant pairings. Visually, it's pleasing seeing different colors, heights, and varieties of plants grown together, but it's nice knowing that there are actually several advantages to doing so. The following are a few beneficial plant pairings.

- **ATTRACT POLLINATORS AND BENEFICIAL INSECTS:** Pairing flowers next to pollination-dependent vegetables can bring pollinators like bees and butterflies to your garden. Nasturtiums have been one of my favorite flowers to add around the garden because of the bright pops of color, the many pollinators they attract, and the fact that they are also edible. They make a good addition to any salad or drink!

- **IMPROVE SOIL QUALITY AND PLANT HEALTH:** Some vegetables, like peas, help to make nitrogen more available in the soil. Nitrogen enhances the growth of other vegetables like carrots, zucchini, and radishes. Plant these two types of vegetables side by side to improve their health.

- **DETER PESTS:** Certain plants act as natural insect repellents and help deter a number of pests from becoming invasive. For example, planting herbs like basil alongside tomatoes can help repel tomato hornworms and aphids (and also provide you with a wonderfully paired dish come harvest time!). Some insects target plants by leaf shape or color, so having a variety of plants can make it harder for insects to determine where exactly their food source is.

- **PROTECT SMALLER PLANTS:** Planting taller plants near smaller plants can provide shade for those needing more protection from the sun. Beans, for example, provide great shade for vegetables like spinach.

- **PROVIDE A NATURAL SUPPORT SYSTEM:** Tall, sturdy plants like sunflowers and corn provide natural support to a variety of different climbing crops, such as cucumbers and squash. This can sometimes alleviate the need for trellises or stakes.

- **SAVE SPACE IN THE GARDEN:** Planting short-season crops like Swiss chard or lettuce in between rows of melons can save space in the garden. By the time the melons are vining and needing more space, you will have already harvested most of your short-season crops.

Companion planting allows you to use natural methods to fight pests, enhance soil quality, and save space while diversifying the ecosystem. It makes your job as the caretaker easier, your garden more visually stunning, and the plants happier!

A GARDEN FOR POLLINATORS

QUILENTHIA WINGFIELD-ACCIME

@mydearestgarden

Quilenthia Wingfield-Accime and her husband, Rodney, are the gardeners behind the YouTube channel and Instagram Dearest Garden. They are plant enthusiasts who enjoy the pollinators, slow living, and food their garden provides them. You can watch their dreamy garden vlogs with a few how-tos sprinkled in along the way online.

Like many others', our garden was born in the midst of the 2020 pandemic lockdown. I had been sewing for several years at the time, mostly for work, but I was in need of a new pursuit that also allowed me to breathe fresh air and escape from the everyday uncertainty. In came gardening.

My husband, Rodney, and I purchased our first home in 2018. I had always wanted a large garden. I dreamed of creating a backyard oasis—a place to escape, a place of magic. And in 2020, that's what I did.

Now, as you walk into our kitchen garden, you are greeted with a trellis of vegetables in grow bags, along with flowers that are known to attract pollinators. We have six black raised beds within the fence border, as well as a cut flower garden. Zinnias, sunflowers, and cosmos are my go-to blooms. I wanted to mix flowers with vegetables to give my garden a potager look.

Having flowers near fruit and vegetable plants is also an excellent way to bring more pollinators into your edible garden. Early on, I learned that this is one of the most important things in a vegetable garden that many beginner gardeners tend to overlook. From the bees to the butterflies, hummingbirds, and everything in between, a garden is not complete unless it is bringing in life. But how do we go about attracting these pollinators to our garden? The simplest and most affordable way is to grow flowers from seed—which brings me back to my go-to flowers. In my experience, zinnias, sunflowers, and cosmos are three of the easiest annual flowers to start from seed that not only attract pollinators but also bring absolute beauty to your spring and summer cutting gardens. Unlike perennial flowers, these annuals only grow for one season and do not come back the following year, so I sometimes like to save my own seeds at the end of the season.

Zinnias, sunflowers, and cosmos come in several different varieties, shapes, colors, heights, and bloom sizes. My favorite thing to do with them is to cut them for flower arrangements. I haven't quite mastered the art of bouquet-making yet, but playing with flowers brings peace to my heart and a smile to my face. I don't cut all of them though. Eighty percent of the flowers stay attached to their stems for the birds and pollinators to enjoy. Watching these creatures freely dance from bloom to bloom is a sight to see, and if you have ever witnessed it, you know exactly what I mean. I look forward to those moments most of all.

On any given day, you can find me puttering and tinkering around the garden. I can honestly say I spend half of my time there. Things begin to slow down in the fall and winter. While I do love to garden in these two seasons, I also like to use this time to reset my mind for the growing season to come.

I feel peace and serenity when I am in my dearest garden. To see the butterflies pour in, to hear the sound of the birds as they wake with the sun, to see my children's faces light up with excitement when they are picking flowers–it has brought so much joy to my heart. When I first began, I thought I would be the one cultivating my garden, but in turn it has been the one cultivating me.

COLORFUL ARRANGEMENTS FROM MY OWN BACKYARD

NICHOLLETTE SHORTS
@broadwaygardener

Nichollette Shorts is an actress with a passion for gardening. She grows a variety of plants, from florals for backyard bouquets, to fruits and vegetables for her family and friends.

Growing up, I can't remember a time that I didn't want to be outside among the birds and trees, or out at the lake with my friends. I spent a lot of time with my grandmother, and she taught me how to care for house-plants at a young age. One summer, she charged me with the responsibility of watering the plants and maintaining them while she was away. I remember feeling so accomplished when she returned home, happy to see her beautifully cared-for plants.

When I moved to New York for college, I had no desire to buy plants or start a garden. My apartment felt dark, as I had only two windows for natural light, and the backyard was too overgrown to tackle, given my busy schedule. I didn't have much of an interest in plants again until I found myself alone at home suffering from depression. I was recovering from a fall and unable to continue the career I had worked so hard for.

This is when I began to realize that all of that unused space in my backyard was a potential garden. At the time, I was limited to what I could do on my own, but what makes gardening so incredible is that you can grow in whatever way fits your lifestyle. Urban gardeners can utilize a variety of container options, and garden beds can be adjusted to a workable height. That summer I started a little container garden full of herbs and cherry tomatoes. By the following summer, my hobby had grown into a full-blown cottage garden in the middle of Brooklyn, New York.

I have since moved to New Jersey, and my garden has become one of my favorite things about our 1920s home. In my garden, I grow a little bit of everything, utilizing raised beds, the landscape, and containers. I enjoy being able to harvest fruit, vegetables, and fresh flowers from my backyard. I love to put together my own freshly grown flower bouquets—so much so that

we added a flower garden to the front of our house to grow flowers for our wedding! I now have a cutting garden in which to create and expand my love of fresh florals by building fun and colorful arrangements.

I go out to the garden each day to calm my mind and reset my day. My favorite thing to do is to take a walk through my garden and note any changes happening during that season, and what potential harvests may be coming my way. Starting a garden and watching it grow has made me feel like I did something right, at a time when everything was going wrong.

CHAPTER 4

GARD

WITH

ENING

KIDS

STEWARDS OF THE EARTH

TENILLE JOHNSON
@puremomblog

Tenille Johnson is an artist, lifelong nature lover, writer, and mom/lifestyle blogger. She gardens with her daughter, Aven Pure Smith, in zone 8a in Georgia, and in zone 13 in Jamaica, at their short-term/vacation rental, CasaLamar Beachside Retreat.

My ancestors were farmers, stewards of the earth, tillers of the land—for survival mostly. My return to the soil is instinctual, intentional, and an opportunity not only to prepare to survive, but also to thrive. As a mother, I share almost everything with my daughter, and I've chosen to share with her my love for gardening. Although it's small scale, our garden is an immense classroom with trees for walls and the sky for a ceiling. We learn so much here. How tiny seeds, given time, proper care, and favorable conditions, soon grow and bloom, providing us with bountiful blessings. Even when conditions aren't favorable and present new challenges, our love for the garden and respect for the process teaches us to problem solve and seek solutions.

In 2019, I started to expand my gardening knowledge. I bought some seeds and put them in pots all across our kitchen floor. My daughter and I watered and watched them grow. I think we were both amazed at the results, and even though there were plenty of fails, the experience was still a win. Then, in 2020, when the world came to a halt, a good friend and I built four garden boxes to expand my home gardening capacity. With self-reliance and sustainability in mind, I planted as much as I could, and I came to understand the value of my practice from the year before. Witnessing tragedy after tragedy and trying to navigate a "new normal" with so much uncertainty, gardening became a part of my meditation. I aligned myself with the pace of nature and spent time with my plants, reveling in the sounds of birds and the smell of basil and mint, instead of feeding anxieties by listening to the latest news stories.

My daughter and I ate breakfast in the garden, used fresh herbs for cooking, and picked romaine and butter crunch lettuces for salads. I even included gardening in my daughter's homeschool curriculum. For her, gardening has provided an opportunity to explore a whole new world right in our

backyard. She gets to develop an understanding of where food comes from and appreciate its journey to her plate. Earthworms and insects like beetles and bees are no longer a source of fear; instead, my daughter is intrigued by their role in our garden. She can tell you how the hummingbirds visit the flowers for nectar and how the bees do the same, spreading pollen and encouraging more of our plants to produce. Watching her learn to handle fragile beings with care, practice consistency, and save her own fruit seeds for planting later has done my soul good.

Gardening with my daughter has had so many benefits, including aiding in her sensory development, providing us an opportunity to bond, and encouraging her interest in eating healthy foods. The experience has also given her a strong sense of accomplishment and taught some valuable life lessons. We've learned the process of trial and error, that sometimes mistakes happen, and that some things are out of our control. We learn to cope in the garden, to try again, and to never give up. We also learn that eventually we will see the fruits of our labor.

I encourage anyone and everyone to work the soil and become a sower of seed. It connects us to Mother Earth, gives us something to look forward to, and brings an abundance of beauty and awe into our daily lives. Start wherever you are, however you can; involve your family, and don't be intimidated; keep in mind that we're all practicing, learning, and growing together.

FROM THE PROJECTS TO HOMESTEADING

DOMONIQUE ROBINSON BROWN
@faithfamilyhomestead

Domonique Robinson Brown runs the YouTube channel Faith Family Homestead. She's a wife and mother of four, living on 2.6 acres in the coastal part of South Carolina, zone 8b.

For the first twenty-two years of my life, I grew up in the projects of New York City. My brother and I were adopted by my grandmother, and we ate the typical American diet, with lots of processed foods, few fruits, and no veggies for me. When I first moved down to the South with a three-year-old son, I did not know how to cook and knew nothing about where food came from. It wasn't until I was met with the challenges of being a single-income family, and I realized that food plays a major role in children's behavior, that I began to explore gardening.

I watched all the documentaries and was convinced that I needed to make a change in our diet to include not just fruits and vegetables, but organically grown produce. With this realization, I also had to figure out a way to do it without spending our whole paycheck at Whole Foods or the farmers' market. I looked into lists like the clean fifteen and dirty dozen to see which produce I could safely buy nonorganic.

In my quest for better food options for my family, I came across a neighbor who was growing 80 percent of their food in their small backyard. At this time, we were living on a quarter-acre property on a corner lot with no HOA. I decided that, instead of buying organic produce, I would grow my own produce for cheaper and get a lot more of it. That's when we turned our whole side yard into a garden.

We are a homeschool family and I am a stay-at-home mom, so it was just natural to include my children in gardening or any project I had going on. I would never be able to accomplish as much as I do without the help of my family. Of course, my children are like any other kids who want to play games and watch shows on their laptops, but I use those activities as a reward. They have learned life skills that they will be able to take with them through their lives, like baking, cooking, caring for animals, canning, and butchering, all of which started in the garden. I am a firm believer in many hands making light work. The faster we get things done, the faster we can move on to things they enjoy more.

Not all my children enjoy gardening, and that is fine, but they all participate in the process. One of my daughters says she wants to be a writer or a farmer when she grows up. Another loves to bake and cook with the produce we grow. My youngest has not found her thing yet, but she enjoys being anywhere I am, and that includes the garden. My son is the muscle—he carries heavy things for me when my husband is at work, and he also cares for our bigger animals. Gardening has become the foundation we never knew we needed.

GROWING UP IN THE GARDEN

ALEXIS BUMPERS
@theurbangardenher

Alexis Bumpers is a certified DC Master Gardener who has been growing food in her backyard since 2018. She was introduced to gardening as a young girl by her grandfather, who grows fruits and vegetables on acres of her grandparents' land in Valdosta, Georgia.

I was introduced to gardening as a little girl when I visited my grandparents in southern Georgia. My grandfather had a plot of land where he grew fruits and vegetables. I started my own garden after my husband and I purchased our first home in 2018. We stumbled upon a property in Washington, DC, that happened to have a large backyard area. On our deck is where I first began growing tomatoes and squash in small planter boxes.

I have an urban garden of fruits and vegetables with pollinator flowers that add a pop of color as well as other beneficial properties, like pest prevention and pollination. I love being able to watch the entire life cycle of a plant, to see how a seed can grow into a fruit or vegetable. Each step of the plant life cycle intrigues me. From the time a seed germinates, to when it produces flowers or leaves, and then when it is time to finally harvest. This process will never get old for me.

Our garden is my sanctuary of peace. I could be having a bad day, feeling overwhelmed, or faced with a challenging decision, and the garden helps to provide my mind with clarity and perspective. It's also my place of solace where I can momentarily escape any anxious thoughts and build myself back up from feeling broken. Watching something grow from a tiny seed that I planted in the ground provides a sense of accomplishment and fulfillment and makes me feel like I have a purpose in life. This feeling is amplified when my harvest provides meals for my family. Gardening has taught me to be patient and appreciate the successes and failures that come with learning to garden.

As it did for others who started gardening for the first time during the pandemic, this pursuit became a healthy outlet to calm any anxieties surrounding COVID-19, and for me it was also a time of becoming a new mom. Gardening with my family and being at home with a new baby during

the pandemic made for some of the most special moments that I will always cherish. We built our backyard to not only grow food but also give us a space to maintain social distance while continuing to see our family and friends outdoors.

My oldest daughter grew up in the garden, as I would carry her in my baby carrier while I was gardening. As she got old enough to walk, I let her participate in more gardening tasks, like digging in the soil with her own shovels, planting seeds, watering plants, and harvesting vegetables. I also gardened during my entire second pregnancy up until a few days before I delivered my second daughter. Gardening has always played a special part in both of my children's lives. Our garden is not only a source of food and beautiful flowers but also a classroom with many lessons and proverbs that can be applied to life.

A FAMILY OF GARDENERS

SHERRIE LEE-WHITE
@theokraladyllc

Sherrie Lee-White is a proud mother of boys, wife of a retired Air Force veteran, gardener, and owner of the Okra Lady LLC in San Antonio, Texas.

We are a family of gardeners. I truly believe that it is never too early to introduce children to the joys of gardening and growing their own food. I learned from my parents, and I'm passing the knowledge on to my sons. I started by allowing them to help me with tasks like watering, weeding, and planting. It has strengthened our family bond and fostered a love for nature and healthy living in my boys.

I love having the opportunity to connect with others in the garden, whether I am working with family members and neighbors, or out in the community. Plus, there is something about the physical act of gardening that is healing. Digging in the dirt, removing grubs, relocating toads, and carrying bags of soil can be a great form of exercise, which releases endorphins and helps to reduce stress. Being outside in the fresh air and sunshine boosts my mood and improves my overall well-being. With each new planting season, I can create something beautiful and unique in my garden. I get to experiment with different color schemes, plant combinations, and layouts to create a beautiful, edible landscape that reflects my individual style and personality.

Gardening with my sons has become a cherished ritual, a shared journey that extends far beyond the confines of our backyard. Their involvement is more than a set of helping hands; it's a collaborative dance of curiosity and enthusiasm. We start with the basics—preparing the soil, selecting seeds, and discussing the unique personalities of each plant. Their young hands eagerly dig into the earth, forming a connection that transcends generations. They water the plants with a mix of serious intent and unbridled joy, understanding the delicate balance needed for growth. We engage in the art of weeding, a task that transforms into a game of spotting the intruders and protecting our green kingdom. Amid the routine, there's always room for surprise, like the time we discovered a curious bunny nibbling on our lettuce, turning an ordinary day into a bunny-chasing adventure. As the garden flourishes, so does their knowledge. They ask questions that lead to impromptu lessons on the ecosystem, the importance of bees, and the magic hidden in a tiny seed. There's a sense of shared accomplishment when the first tomato appears or the sunflowers reach for the sky. It's in these moments that I see the seeds of responsibility and care taking root in their hearts.

Our gardening escapades are not just about plants; they're a canvas for laughter, storytelling, and the simple joy of being together. In the canvas of our backyard, memories blossom alongside flowers, creating a tapestry woven with the threads of love and shared growth.

USDA PLANT HARDINESS ZONES

You may have heard of or read the term *plant hardiness zone* before, maybe on the back of a seed packet, but what exactly is it? Plant hardiness zones are determined by your region's lowest annual temperature; they help define which plants will be able to survive and flourish best in your area. Becoming conscious of these zones will save you a lot of time and money in your gardening.

The easiest way to find your zone is by searching for your zip code and "planting zone" on the internet. The zones range from 1 to 13, with zone 1 having the most extreme low temperatures—like Alaska—and zone 13 being the warmer, tropical regions. The zones are calculated in Fahrenheit, and each zone is either 10 degrees warmer or colder on average than the one adjacent to it. After experts did more research and studied weather patterns over a longer period of time, they adjusted the map to show subzones as well, which are calculated in 5-degree increments so that planting can be more precise. The subzone is listed at the end of the zone number as either a (the colder part of the zone) or b (the warmer part); for example Charlotte, North Carolina is in zone 8a, while Myrtle Beach, South Carolina, is in zone 8b.

Knowing this information is most beneficial to gardeners growing perennial plants, trees, or shrubs; unlike annuals, these plants are meant to live for much longer than just one year. What enables a perennial to grow back every year is its ability to withstand the highest and lowest temperatures year-round, so it's important to become familiar with the climate in your area. For example, the mango is a tropical fruit tree that can survive outdoors in only a few places in the United States, like Hawaii, Southern Florida, and Southern California. If you were to try to plant one of these in the ground somewhere in New York or Alaska, the extended exposure to the cold temperatures would likely kill it in its first winter season.

Planting species native to your area is the best way to have a stable, healthy garden or yard. Native plants, also referred to as indigenous plants, are plants that naturally grow in certain regions and habitats and have formed a relationship with the land over thousands of years. They usually require very little maintenance from us and are almost guaranteed to thrive, as they've adapted themselves to the climate and conditions for that area. When you're shopping around for plants, especially perennials, check the tag for the zone range. It's almost always listed and can help you decide which plants are suitable for permanent landscaping in your garden.

Some plants may not survive all year round, but it doesn't mean you can't still enjoy their blooms, even if it's just for a short season! The USDA Plant Hardiness map should be considered a guide. You should feel free to experiment with a variety of plants, especially if you have a greenhouse or plan to plant them as annuals. Remember, your garden is your own science experiment; explore what works for your own space.

PLANT HARDINESS ZONE MAP

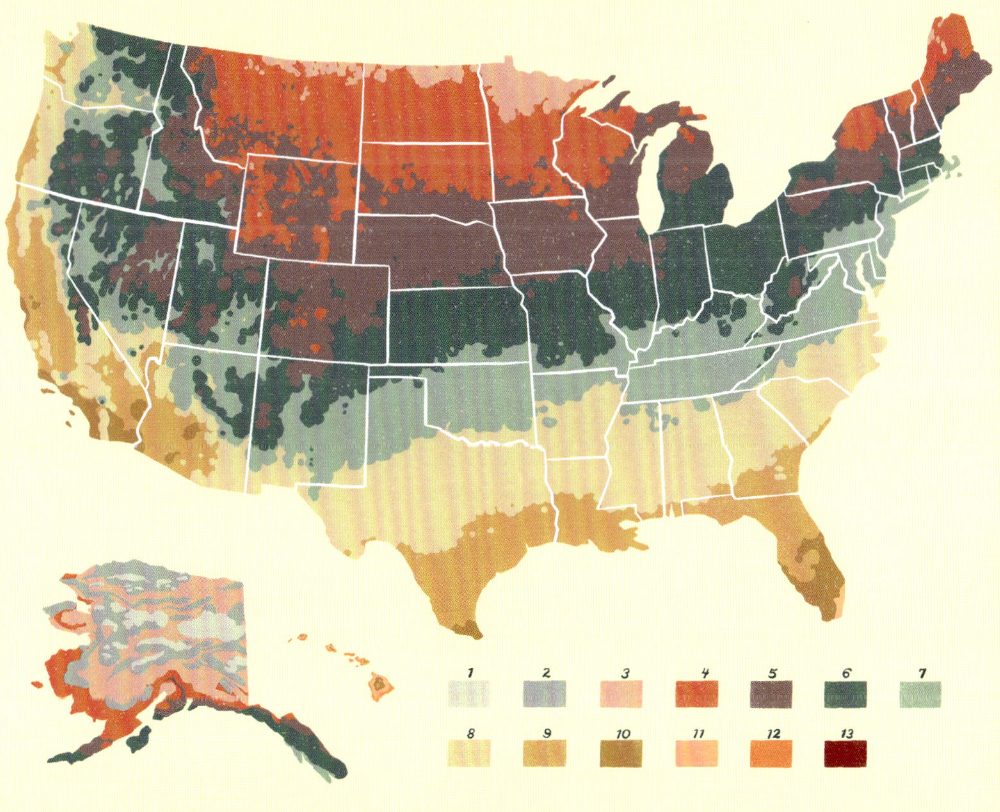

FULFILLING, INTENTIONAL LIVING THROUGH GARDENING

TONETRA SMITH
@princessnetraa

Tonetra Smith is a small-space backyard gardener located in San Antonio, Texas, and a mother of three boys, two of them twins. She teaches beginner gardeners how to garden through social media platforms including YouTube and Instagram.

When I was twelve, my mother was accepted into medical school at "Nature Island," better known as Dominica. The youngest of four siblings, I would be the only child moving. I felt so lucky; I knew living on an island was a once-in-a-lifetime opportunity. I was ecstatic to leave a very urbanized country to take on the adventures of cultural island living.

I arrived in awe of the lush green tropical rain forest that blanketed this beautiful oasis. Our Dominican neighbors lived simply: They grew their own food, raised animals, and cooked from scratch. Little did I know this was the beginning of my love for gardening!

Fast-forward to January 2020. I had just become a new mom, and all I wanted to do was stay home and care for my son. I had been a barber at our family business for the previous ten years. I was burned out! I was yearning for peace and fulfillment in my life, and so tired of what felt like a mundane, unfulfilling lifestyle that I had created. Then the pandemic hit, throwing me out of work—which, considering my feelings at the time, wasn't the worst thing. I recognize my privilege in this regard, as so many people were severely impacted by the pandemic. It did, however, drive me into survival mode as a mother and protector of my son, and brought on a sense of uncertainty that unsettled me to my core. I knew that I needed to provide, and the most essential way was food for my family. Right then and there, that twelve-year-old girl who had traveled to the Dominican Republic was reborn. My love for nature flooded in, and my gardening journey began.

My small backyard garden is my sanctuary. As it's situated in a very urban area filled with life and noise, you might be surprised to hear sounds of buzzing bees and purrs of outdoor cats, and to see views of ripe red tomatoes and vibrantly colored flowers

THE NOTORIOUS
Brooklyn's Finest

throughout my garden. When I'm gardening, all the busy noise seems to fade away, and it's just me and Mother Earth at work.

When it's time for a little extra excitement, I'll bring my boys out to get a dose of nature and water their very own garden! My kids love gardening so much; they are strong, helpful, and eager to learn. Swinging in the garden is one of our favorite pastimes. Feeling the fresh breeze while looking at the night sky and hearing the sounds of nature is calming for them.

Gardening has brought my family immense joy. It's become a lifestyle for us. I feel the most fulfilled and intentional when gardening with my kids. We've had so much fun learning where our food comes from and how it grows. Knowing that we have the power to grow our own food brings us a sense of security. The process of gardening has also instilled core values in my children, like the importance of family, creativity, passion, and gratitude. Our garden is ever-changing and evolving, much like we are.

Inspired by the impact gardening has had on my family, my mission is to inspire fulfilling and intentional living through gardening. Children are a huge part of my mission, because kids are the future! They also happen to gravitate to me, and I welcome them with open arms. Kids have a way of making life not so serious, and I think that's what a lot of adults need. I can teach children the importance of community, leadership, respect, and happiness, all through gardening. It's so amazing how brave and accomplished a child feels after planting a tomato and watching it grow! I enjoy doing volunteer gardening work at my son's school, and I also give back to the community by growing and giving free plants at public events and sharing my knowledge and inspiration wherever I go. If I can teach just one person how to grow their own food, I will have done my duty. So I've set out on this lifelong mission to teach as many as I can.

$1.44
Southern Charm
CUCUMBER
AT

LESSONS ON WHERE YOUR FOOD COMES FROM

ESTELLE NGOKOLO
@myprettyfem

Estelle Ngokolo is a mom of three little children and lives with her family in the beautiful German region of Bavaria.

When I was a few months pregnant and our first daughter was already two, my husband and I moved into a larger house. For us it was very important to have space, and above all a garden, no matter how big, for our children. We wanted to give them a certain kind of childhood, where they could be in contact with nature and learn to respect, value, and appreciate what it has to offer. The day we decided to clean up a little corner of the garden is still engraved in my mind; it feels like it was yesterday. It was my project. I didn't have a plan, and my husband didn't know anything about gardening—he just wanted to help me realize what I envisioned in my head. All I knew for sure was that I had a two-year-old daughter who didn't like potatoes, carrots, and especially fresh tomatoes. I hoped that if I showed her how to plant these vegetables and how they grow—if she sowed her own carrots, watered them, and watched day by day as they grew—she would end up eating them and enjoying them. And guess what? It was a success! My daughter started to eat the fresh tomatoes from the garden like cherries, and she began to enjoy potatoes too. She learned how these vegetables are planted, and that they need sun and water and, above all, time to grow.

When I started gardening, I had three main motivations. First, I wanted to teach my children to eat vegetables; second, I wanted to offer my family organic food; and third, I felt nostalgic about gardens, because I grew up with one. But these past years since I've planted my garden, I've developed a very strong fourth reason: Gardening has become a therapy, an hour in the evening in the quiet, looking after my plants. Motherhood can be very stressful and isolating (which is one reason why I opened my Instagram account, to connect with moms around the world. Having a garden, where I can sit in the fresh air, surrounded by my organic vegetables, and listen to my children play and have fun, is so calming. It's one reason why we decided to live in the countryside, and why I do the gardening myself, with the help of my husband and our children.

If I have any advice to give to mothers, parents, or new gardeners, I'd say that together with your children, plant or sow what they don't like to eat. Also, plant what *you* like and use a lot in your cooking, because believe me, organic tomatoes from your own garden are nothing like those from the supermarket—they're so much better! You don't have to know everything from the start. Each season will teach you a new lesson, so never compare your garden with someone else's. Gardening is a gradual process. Little by little, you will get there.

Prüf Nr
ZTV-SA K1

CHAPTER 5

COMM
GARD
AND FA

UNITY

ENING

RMING

SISTAS IN THE VILLAGE

BWEZA ITAAGI
@bweza
@sistasinthevillage

Nyabweza "Bweza" Itaagi is an urban farmer, community cultivator, and horticulturist. In 2015 Bweza moved from Colorado to Chicago to pursue a master's in sustainable urban development at DePaul University. Bweza co-owns the SITV farm and is the steward for the Englewood Nature Trail and Agro-Eco District.

In the spring of 2020, I cocreated Sistas In The Village (SITV) farm alongside my soul/soil sister, Mecca Bey. We felt a clarion call to grow food for our community, in the heart of the pandemic. In our first two years, we participated in an extensive farming apprenticeship program led by Erika Allen, cofounder of Urban Growers Collective. This program was foundational for us to learn how to work with the land and maximize production on smaller city plots, and it gave us great guidance on how to formalize our business as farmers. Today, SITV has a permanent home in the heart of the Englewood neighborhood on the South Side of Chicago. We are committed to growing food to nourish an intergenerational African Indigenous community, with a focus on crops central to the African diaspora. On about half an acre of land, we are growing vegetables, fruits, berries, herbs, and grains for our neighborhood.

At SITV, we're reclaiming farming as a source of cultural pride and spiritual resilience. We envision a community in which we are all fed spiritually, physically, and emotionally. We sell and donate the food we grow, host workshops for community members of all ages, and partner with like-minded organizations and brands. We have an ongoing relationship with a local Girl Scout troop where we teach the girls and their families how to grow and prepare food from seed to table, and we host story circles and knowledge exchanges on the land. This year, in partnership with herbalist and medical student Sharmain Siddiqi, we started a community apothecary, Sister Apothecary, with free plant medicine for our neighbors. The model is take what you need, leave what you can.

It's almost impossible to express how beautiful this journey has been. The farm has been central to my own healing journey and

SISTER
APOTHECARY

a space where I've experienced profound care and support from our village. We've been abundantly blessed with support, and we've met amazing people in the process. I have shed so many tears, had countless spontaneous dance parties, and watched the love of the community come together on our land. Prior to our farm, the site had been vacant for decades, as is the case with vast amounts of land in the greater Englewood community. My hope is that SITV can be an example and an ongoing practice of how to pour love, life, and abundance into spaces vacated by systemic neglect and racist city planning policies. In a short span of time, we've seen the vital parts of the ecosystem return to the land. We're continuously observing different species of birds, insects, and native plants coming home. It's truly a blessing to be able to do this work!

POWER
TO THE
PEOPL

A SPACE OF HEALING AND CONNECTION

BRITTANY LEAVITT
@bleavitt8

Brittany Leavitt is a Community Dreamer who focuses on outdoor advocacy for locations ranging from local parks to wilder hiking and climbing regions throughout the US. She is the executive director of the national organization Brown Girls Climb. Brittany is also the garden coordinator for the Cooper Street Community Gardens.

I was the collector of earthworms, ladybugs, and roly-polies. I was the observer of the milkweed as I watched the monarch caterpillars chomping on the leaves. I was the kid making my own bouquets, who liked to spend time outside watering with my family. Growing up, I spent time gardening with my grandma during the summers in upstate New York, and with my parents at home in Maryland.

Once I moved to New York City, I had to readjust to a lot of things. I had to review how I connect with nature and third spaces. I had known about community gardens but was never active in any until 2022. The history of community gardens is so important. The vision for community gardens began in Detroit during the economic recession of 1890. Community gardens were created by turning old vacant lots into spaces where communities who were dealing with unemployment and food insecurity could grow together while sharing space and resources. And they are still relevant today: We saw a big rise in interest in community gardens in 2020 during quarantine, when folks could come outside and work together on growing foods for themselves and the community. Community gardens can show up in many different forms. Some spaces require you to sign up for a plot, pay dues, and navigate guidelines that may not feel as welcoming as more loosely organized gardens.

I am very grateful for connecting and building a relationship with the garden that I am the garden coordinator for in Brooklyn, New York: Cooper Street Community Garden. The space holds more than just beautiful tall sunflowers and growing tomatoes in abundance; it's a *third space*, where folks can connect, heal, create art, educate each other, and more. It's a space that is intergenerational and mindful, where although you are in the city, you can get lost in observing the bumble bees pollinating the flowers and the snails slowly moving over giant leaves, or hearing the sound of jazz played live by community members or spoken word shouted to the treetops. The community garden is a space where many hands, voices, and thoughts support the growth of food for each other and teach each other radical care. It's a space to close your eyes, take a few deep breaths, count to three, and settle yourself.

A COMMUNITY THAT SOWS TOGETHER, GROWS TOGETHER

CARLENA JENKINS
@earth_n_magic

Carlena Jenkins, known by many as Umi, is a resident of Biloxi, Mississippi, where she serves her community as co-executive director of the Mississippi Farm to School Network, which helps get locally grown food into school cafeterias. She is a proud mother of three children and actively advocates for minority farmers, land retention, and just and equitable food systems across the South.

Since 2013, gardening has brought me immense satisfaction, from spending time with my elders at the community garden to preparing fresh produce for meals that nourish my family. My journey began when I searched for wholesome activities for my three children, whom I was homeschooling at the time. After we'd visited all the local museums and parks, and made frequent trips to the library, I wanted to provide them with more fulfilling experiences, and skills they could reinvest into their community. I searched community gardens online and found 34th Street Wholistic Gardens & Education Center in Gulfport, Mississippi. I contacted the garden director, a Black man named James Franklin, who invited us out and helped us get started with our first square-foot garden bed. It wasn't long before we had four beds growing tomatoes, peas, greens, peppers, carrots, eggplant, okra, watermelon, sweet potatoes, and all sorts of herbs. My children were alongside me, learning about soil quality, planting and harvest times, and preserving produce. Soon we met other Black farmers and growers passionate about uplifting Black agrarian foodways and traditional growing practices. Growing food was and still is an enriching experience, one that I believe in my soul is making our ancestors proud. We are stewards of the land, reviving valuable work by learning how to grow sustainably, organically, and with the community at the forefront.

I had no idea how much the act of digging my hands into the soil would restore my mental and physical equilibrium. The environment we were creating was nothing short of beautiful; in fact, it inspired me to start growing on my small apartment patio, which my family jokes is at capacity. It isn't always easy on hot days—yes, there are insects, and some crops fail while others thrive—but overall, it's the most empowering journey I've embarked on alongside motherhood and my personal relationship with the Creator.

The community that sows together, grows together—this is how we become rooted in food sovereignty. I've maintained long-standing relationships with people from all walks of life who find peace in a common place . . . the garden. The members of my community garden gather to eat and share stories, exchange seeds, and plan for the seasons ahead, reinforcing the need for communal spaces. We may not always have big-name supermarkets to shop in for fresh food—some communities don't even have grocery stores—but I've witnessed how we can lean on one another for inspiration and solutions to food insecurity. I'm thankful my children have grown up with memories of eating muscadines straight from the vine, munching on spicy mustard greens, and plucking figs to add to their toast drizzled with local honey. These experiences and the memories they conjure bring me hope for the future—and as long as Black women are in the garden, what a bright future it is.

ORGANIC FERTILIZING

I admit that I went my entire first growing season without using any fertilizers. I remember how excited I was when I was able to harvest quite a few veggies that season. Now that I have incorporated regular fertilizing practices in my garden, I've noticed how much longer my plants survive, how much bigger they get, and how much better my harvests taste. If you invest a little time in your garden's long-term health by adding in some organic fertilizer here and there, you will be rewarded for years to come. Remember that healthy and balanced soil is the basis for a healthy and balanced garden!

The idea of fertilizing can sometimes be overwhelming. Mostly because it's hard to determine what your plants need, what you should use, how often you should fertilize, and when. Try not to overthink it. The bottom line is that while your plants are growing they are taking nutrients from the soil; over time, the constant sequestering of nutrients leaves the soil depleted. That's when we need to intervene and add organic fertilizers—natural products used to improve your soil's health and restore the nutrients that your plants depend on. The three primary nutrients that plants need to survive are nitrogen, phosphorus, and potassium. Nitrogen supports plant color and growth, phosphorus is responsible for production of fruits and flowers, and potassium helps promote strong roots. Each bag of fertilizer bears a set of three numbers—for example, 5-3-3. This means that the bag contains 5% nitrogen, 3% phosphorus, and 3% potassium.

Plants that yield large harvests usually come from a garden that has healthy, well-balanced soil. I like to think of soil as a human: If it is fed, watered, and given a bit of love, it will be able to produce its best work. Soil is the base of every growing thing and plays a huge factor in how big a yield your garden will produce.

There are many different products that can be used to enhance the quality of one's soil without using any harmful chemicals.

- **COMPOST:** Organic material can be added to any plant in the garden to help deliver nitrogen, phosphorus, and potassium to the soil. Mixing in a layer of compost in the spring is ideal, as it gives the soil a boost of nutrients in preparation for planting. Doing it again in the fall increases the amount of nutrients the plants draw down into their roots before winter, which will improve the quality of their blooms the following spring. This is why it's a great idea to be composting at home. You can turn your kitchen scraps and garden waste into soil to grow new produce, which will then become future meals. Complete self-sustainability! See pages 90–91 for more on composting.

- **FISH EMULSION:** Fish emulsion is a liquid concentrate made from fish parts that needs to be diluted before adding it to your garden. It is high in nitrogen and great for growing bigger, greener leaves. It works well, but it does have quite an odor, so keep that in mind.

- **WORM CASTINGS:** Word castings is just a fancy term for worm poop! This is a slow-release fertilizer beneficial to all types of plants, even your houseplants. Worm castings are full of nutrients that will help your plants grow stronger and give you more plentiful harvests. It's also great at amending soil, as it creates aeration while simultaneously boosting water retention.

- **MANURE:** Animal waste from herbivores makes a great soil amendment. It's full of organic matter and all the things needed to boost your soil's health. Chicken, cow, goat, sheep, bat, and horse manure should be used after composting or aging. If it's fresher manure, you can still add it to the garden, but only in the fall to prep the area for spring planting, giving it time to age. This is yet another great way to make your garden even more self-sustaining.

- **BONE MEAL:** Bone meal is made from animal bones (typically from cows) that are steamed and then ground into powder or "meal." It provides a great source of phosphorus for your veggies and fruits.

- **BLOOD MEAL:** Blood meal, a byproduct from animal slaughterhouses, is essentially dehydrated blood, made into a powder that supplies a large boost of nitrogen to your garden.

- **FERTILIZER BLENDS:** You can also purchase fertilizer blends such as Garden Tone or Plant Tone (my favorites!), which have a mix of active ingredients like alfalfa meal, poultry manure, feather meal, and bone meal. It can be applied monthly during the growing season and acts as an all-in-one fertilizer. If you don't feel comfortable using fertilizers with animal parts, there are other options, like seaweed fertilizers, but they do have lower amounts of nutrients available. You'll need to apply them more frequently throughout your growing season, but they are a viable option to be sure.

BLACK THUMB FARM

ALEXYS ROMO
@blackthumbfarm @we2mustgrow

Alexys Romo is the founder and executive director of Black Thumb Farm, a nonprofit serving BIPOC youth in the San Fernando Valley of Los Angeles through mentorship, leadership training, and hands-on skills on their urban farm. She is a wife and the mother of three children and seven chickens.

I come from families that are very much tied to the land, but you wouldn't know it from the way I grew up. I was born in Kansas City, Missouri, but my family moved to the San Fernando Valley of Los Angeles when I was three. I grew up mostly in apartment buildings with no green space, attending public schools whose playgrounds were covered with asphalt, in urban settings dominated by concrete. The nature in close proximity seemed inaccessible to families with working parents, limited income, and lack of community resources. But that setting is not my heritage. My ancestors were sharecroppers and worked the land for generations. My grandfather was born in Mississippi and grew up picking cotton, shucking beans, and roaming outdoors until (and past) sundown. My grandmother comes from Arkansas, where she grew the best plums and made the best plum jam. At some point, capitalism, oppressive systems, and white supremacy all had a hand in severing the ties to the land that my grandparents knew.

At nineteen, I left the US for the first time ever to volunteer in the Philippines at an orphanage. I was then playing college basketball and was going to coach a youth league there. The morning after I arrived, I was woken up before sunrise by one of the caregivers to go and help prepare breakfast, a porridge made of cassava. I had no clue this would include cutting down, digging up, washing, chopping, and cooking the cassava! It was my gateway into reconnecting with the land. Over the next several months of living there, I learned how to care for fruits and vegetables and chickens, and to cook nourishing meals from the food we had grown. This experience permanently altered my life. Once I was reconnected to the land, I zeroed in on making sure I could make a way for people like me, particularly young people like me, to make these connections too: people who come from ancestors rich in memories of barefoot play in fields, dirty hands from growing gardens, and scents of their mothers and mothers' mothers cooking in the kitchen.

In 2020, I started a nonprofit called Black Thumb Farm with the intention of getting BIPOC youth more connected to their food, their ancestry, their communities, and the natural beauty around us. Since its inception, we have been able to serve hundreds of youth through our programs, host cultural and seasonal events for the public, and grow culturally relevant and important foods on our urban farm. We mentor our youth, provide leadership training, and teach them valuable skills while working on the farm and at our school garden sites. Communing in nature while growing my own food radically healed me in many ways and brought a new light to my life. I've been able to see how participating in farming and gardening can heal others as well, in ways we don't always expect or understand. My three children are being raised to value our earth, our foods, and our ancestors, and we are blessed to do so in a community with many many others who value these things as well. What a gift!

NESTLED ON A BUSY STREET CORNER IS A GREEN OASIS

TAJI RILEY
@unboundbx

Taji Riley is a gardener with roots planted in the Bronx, New York. She learned houseplant care from her mother and gardening skills from her grandmother.

Gardening for me began with my grandmother, Mary Lee Johnson. Grandma Mary became a member of the Willis Avenue Community Garden in our neighborhood upon its opening in 2015. Every summer for nearly a decade I watched my grandmother plant various crops that would feed our family throughout the winter. In our 6 by 4 foot raised bed we grew cabbage, cucumbers, collard greens, peppers, squash, tomatoes, and okra—staple crops because of our heritage and connection to the South. After a few years of volunteering, Grandma Mary became a steward at the community garden because of how involved she was and her adept knowledge of plants and gardening. She helped maintain the common areas, started a simple composting system, and always welcomed fellow Bronxites into this green space. I am from a lineage of women who farmed in order to barter, make money, and feed their families and the community. Grandma Mary taught me everything that I know about gardening, as did her grandmother and great-grandmother. She would tell me stories about watching her mother strip collard green leaves from the stalks and how good she felt about keeping those practices within our family. Community gardens bring such incredible value to our neighborhood, beyond just plants and produce.

In my neighborhood of the South Bronx, a vacant-lot-turned-trash-dump was converted into a community garden. The Willis Avenue Community Garden is a part of New York Restoration Project's fifty-two green spaces throughout the city. NYRP'S objective is to collaborate with residents in communities across the five boroughs to renovate gardens, rescue parks, and promote urban agriculture while building green spaces that transform the city landscape. The garden's beautiful 9,063 square foot expanse features twenty-four raised beds, a butterfly garden, and performance space known as the casita. The community garden is a public green space where all

BRONX
COMPOST
2022

people in the community can feel safe and welcomed. In my eyes, it's the heart of our neighborhood because of its popularity for neighborhood celebrations, entertainment, and education.

Having access to the community garden has increased physical activity among all age groups in a neighborhood where children are twice as likely to be hospitalized for asthma as elsewhere in the city. It is the lungs of the neighborhood because of the additional fresh oxygen the plants give off in this highly congested urban area. These same children have access to a beautiful educational space where they can learn about how food is grown, identify insects, and be outdoors safely. The garden gives a voice to senior adults whose voices tend to go unheard. They can play card and board games, share stories, and grow vegetables in a tranquil environment while connecting to their history and culture.

Nestled on a busy street corner, this green oasis removes me from the chaos of the city and brings me to the present moment. I've learned resilience by growing collard greens in the heart of The Hub, a commercial center in South Bronx. I am learning to trust myself when I plant the crops in early spring and have no clue what will begin to grow in mid-summer. I am cultivating self-trust throughout the process of gardening. Being a member of the community garden is not easy; it is labor, but a labor of love. My family legacy of urban gardeners will live on in my practice of gardening. I hope to expand one day beyond the walls of my community garden, but for now, it is the place that my vegetables and I call home.

THE GREATEST CROPS ARE GROWN IN THE LOCAL COMMUNITY

DOMINIQUE KINSLER
@pharmunique

Dominique Kinsler is a wife, daughter, sister, Florida A & M University pharmacy graduate, and proud member of Delta Sigma Theta Sorority Incorporated, who fell in love with gardening during the COVID-19 pandemic. Her passion for gardening centers on educating the community and embracing her ancestral heritage of farming and agriculture.

At the height of the pandemic in 2020, I began having vivid, detailed dreams where I was surrounded by nature and beauty. When I awoke, I would have a feeling of entrapment as I remembered that the world was completely locked down. I spent my days working from home, oppressed by the weight of my corporate job. The oppression anchored me in depression and anxiety. Despite my feelings, I tried to maintain a sense of normalcy, but night after night the dreams became increasingly vivid, rooting themselves deeper and deeper into my mind. Eventually, I willed my body and hands to follow my dreams. I began to take a closer look at the houseplants I was struggling to maintain, and, channeling my ancestors, my dreams and desires transitioned to the cultivation and growth of my garden. My ideals developed full circle into the realization that in nature, I could nurture and grow so much more from the earth and within myself.

In 2022, shortly after moving into a new home with my husband, I began documenting my gardening journey. As a Black, millennial, educated young woman, I was exposed to a space that my peers were absent from a decade prior. My conscious choice to work in my garden has heightened my realization of privilege. I was and am gardening as an act of rebellion and reconnection to my roots in my beautiful home. This new awareness encouraged me to pursue community gardening. I began to pivot to assisting family, friends, and other disadvantaged individuals with the development of community gardens in my local area. Driven by my desire to serve and support, I continued to explore, and while doing so, I discovered a hidden gem: a Black-owned community garden in the heart of downtown Orlando. This discovery was unlikely, since it was located directly next to what is known as a government-owned housing project, commonly dubbed "the hood," "the ghetto," or "the projects." While serving in this garden, I made wonderful discoveries, but the most memorable sighting was the most beautiful collard green plants I had ever seen.

As I was overwhelmed with a feeling of pride, I came to the realization that my vivid dreams were now a reality and my labor was now love; all of my work was now a manifestation of both my ancestors–specifically, my grandfather–and my wildest dreams. Paul Fleischman shared, "Community gardens were oases in the urban landscape of fear, places where people could safely offer trust, helpfulness, charity, without need of an earthquake or hurricane . . . Community gardens are places where people rediscover not only generosity but the pleasure of coming together."[1] I salute all those who give their time and talents to rebuilding that sense of belonging. While my online followers may see gardening as an individualistic and therapeutic activity, the greatest crops are grown in the local community. As I continue my gardening journey, the community will remain at the center of my focus.

---

1 Paul Fleischman, *Seedfolks* (HarperTrophy, 2004): 48.

ORGANIC PEST CONTROL

Being an organic gardener is about growing plants in tune with natural processes, ecosystems, and our bodies. We are drawn to the colors, smells, and delicious foods our gardens produce, but we aren't the only ones. Bugs, deer, and rodents are also attracted to healthy looking plants; don't take it personally when they want to indulge in your beautiful veggies just as much as you do! Unfortunately, "pests" are inevitable when we're growing in their habitat, but it is only a matter of perspective. Remember that in cultivating outdoors, we are in their home. Be a good guest! Balancing patience with vigilance is key to a flourishing organic garden where all can coexist together. Still, there are *some* things you can do to stave off pests from your garden.

- **GIVE YOUR PLANTS SPACE AND VARIETY:** One of the easiest ways I've found to keep diseases from spreading rapidly is to make sure your plants have proper spacing between them and to not plant too many similar things together. Certain plants are susceptible to the same diseases and pests; if they are all planted together and one gets infected, that disease or pest will spread more easily and potentially kill an entire section of your garden. I like to use the methods of companion planting and polyculture to help naturally assist in the prevention of unwanted pests. These methods are similar. Polyculture involves growing more than one type of crop in a designated space; companion planting involves placing a variety of certain plants together to help with pest control, pollination, soil quality, and increase crop productivity (see page 138 for more details on companion planting).

For example, instead of growing all of my cucumbers or melons in one garden bed, I space them out throughout the garden, planting a variety of flowers and other vegetables in between. This can make it harder for pests to locate their food source.

- **PROTECT WITH GARDEN NETTING:** Some type of simple barrier such as insect netting or row covers is another great preventative option. These let in sun, water, and air but keep out caterpillars, slugs, aphids, and any other critters that may want to chew on your plants. It's important to put the netting or barrier down when the seedlings are young, since this is just a preventative measure and not a treatment. If the pests are already on your plants when you lay down the netting, you'll just be trapping them inside, giving them free access to all of your veggies. For fruit bushes, you can bag the fruits in little mesh bags before they become ripe, helping to keep the birds from making these one of their favorite snacks.

- **USE NATURAL PEST SPRAYS AND FERTILIZERS:** Of course, sometimes no matter how hard we try to discourage pests, they still manage to find a way in. If you've found that you are now having to treat your plants, there are a variety of organic pest control solutions available so that you never have to use any harmful pesticides. When we use products with chemicals in the garden, they kill not only unwanted visitors but also pollinators like bees, butterflies, and moths, as well as beneficial insects that feed on the pests you're trying to control. Furthermore, research shows that produce treated with pesticides actually produces fewer beneficial compounds for our health than produce grown organically. Since plants create their own constituents like antioxidants and alkaloids to prevent pest attacks, coddling the plants with pesticides can make them produce fewer beneficial compounds. There are a variety of organic sprays you can use that will not kill pollinators or affect the nutrition of your harvests.

- **DIY NATURAL PEST SPRAY:** Mix biodegradable soap like Dr. Bronner's with water. You can also add a couple of tablespoons of cayenne pepper or garlic to this mix to make it more potent. This solution helps kill some of the most common pests, like aphids, thrips, and mites. Spray directly onto the plant and its leaves early in the morning (never in direct sunlight), and rinse the leaves with water once the soap dries, to prevent burning. I've found this to work well in the garden, as it doesn't kill any of the beneficial insects that I need for my garden to survive.
- **ORGANIC FERTILIZERS:** Maintaining healthy soil can prevent a lot of plant diseases. Try using fish meal, worm castings, compost, blood meal, or chicken manure. (See pages 208–11 for more details on organic fertilizers.)

There will be a lot of trial and error in figuring out what techniques work best for the type of plants you have in your garden. With all the prevention in the world, even the most experienced gardeners will still be dealing with half-eaten leaves or having to pick hornworms off their tomato plants.

The more you stay on top of daily maintenance by removing any fallen fruits and other dead plant debris, the easier pest control will become. However, as I said before, approach these pests with a soft heart. They are the same as all of us, just out here trying to eat!

CONTRIBUTORS

ALEXIS BUMPERS
@theurbangardenher

ALEXYS ROMO
@we2mustgrow
@blackthumbfarm

ALNISSA RUIZ-CRAIG
@alnissa.grows

ALYSSA WARD
@lyssacouture
@wardsfarmnj

ASHLIE THOMAS
@the.mocha.gardener

ASIA SPRATLEY
@yellowdoorurbanhomestead

BATAVIA CHOCTAW
@b_bettagarden

BRITTANY LEAVITT
@bleavitt8

BRITTNEY HUTCHERSON
@themelaninurbangardener

NYABWEZA "BWEZA" ITAAGI
@bweza
@sistasinthevillage

CARLENA JENKINS
@earth_n_magic

CHRISHA FAVORS
@naturally_chrisha

CIEARRA EVANS
@thethriftedplanter

DOMINIQUE KINSLER
@pharmunique

DOMONIQUE ROBINSON BROWN
*@faithfamilyhomestead_*

ERIKA DUDLEY
@erika_francoitalian

ESTELLE NGOKOLO
@myprettyfem

HANNAH VEGA
@afroforagers

JANICE GROVES
@jaysgardenjournal

JULIA BENN
@julbeartgardens

KOLA STARNES
@luvya_garden

LEOTA WILSON
@curlycultivators

NAIMAH ABRAHAM
@nimzyunmeasured

NICHOLLETTE SHORTS
@broadwaygardener

QUILENTHIA & RODNEY WINGFIELD-ACCIME
@mydearestgarden

SHAQUITA JACKSON
@thegreendesert

SHERRIE LEE-WHITE
@theokraladyllc

TAJI RILEY
@unboundbx

TENILLE JOHNSON
@puremomblog

TONETRA SMITH
@princessnetraa

URSULA CARMONA
@carmonaacres
@homemadebycarmona

PHOTO CREDITS

5: Photograph by Kyle Grossman

6: Photograph by Amber Grossman

11–12, 15, 18: Photographs by Kyle Grossman, featuring Amber Grossman

19: Photograph by Amber Grossman

21–25: Photographs by and featuring Ashlie Thomas

26, 28: Photographs by Jose Ochoa, featuring Shaquita Jackson

29: Photographs by Hanif Burt, featuring Shaquita Jackson

31–35: Photographs by and featuring Ursula Carmona

36, 38–41: Photographs by and featuring Leota Wilson

44: Photograph by Amber Grossman

45: Photograph by Kyle Grossman, featuring Amber Grossman

47–51: Photographs by and featuring Chrisha Favors

52, 54–57: Photographs by Ashley Neuworth, featuring Hannah Vega

59–61: Photographs by and featuring Asia Spratley

62, 64–67: Photographs by and featuring Ciearra Evans

72: Photograph by Amber Grossman

73: Photograph by Kyle Grossman, featuring Amber Grossman

75–77: Photographs by and featuring Erika Dudley

78, 80–83: Photographs by and featuring Naimah Abraham

85–89: Photographs by and featuring Brittney Hutcherson

91–92: Photographs by Amber Grossman

94, 96–99: Photographs by and featuring Kola Starnes

101–105: Photographs by and featuring Batavia Choctaw

106, 108–111: Photographs by and featuring Alnissa Ruiz-Craig

114: Photograph by Kyle Grossman, featuring Amber Grossman

118–119: Photographs by Amber Grossman

121–125: Photographs by and featuring Alyssa Ward

126, 128–131: Photographs by and featuring Julia Benn

133–137: Photographs by and featuring Janice Groves

140, 142–143: Photographs by and featuring Quilenthia Wingfield-Accime

145–147: Photographs by and featuring Nichollette Shorts

150: Photograph by Amber Grossman

151, 153–155: Photographs by and featuring Tenille Johnson

156: Photograph by Sarah Canfield, featuring Domonique Robinson Brown

158–159, 161: Photographs by and featuring Domonique Robinson Brown

160: Photograph by Sarah Canfield, featuring Domonique Robinson Brown

163–167: Photographs by and featuring Alexis Bumpers

168, 170–173: Photographs by and featuring Sherrie Lee-White

177–180: Photographs by and featuring Tonetra Smith

181: Photograph by Kimberly Mccall

182, 184–187: Photographs by and featuring Estelle Ngokolo

190–191: Photographs by Amber Grossman

193–197: Photographs by and featuring Nyabweza "Bweza" Itaagi

198, 200–201: Photographs by and featuring Brittany Leavitt

203–207: Photographs by and featuring Carlena Jenkins

210: Photograph by Kyle Grossman, featuring Amber Grossman

212, 214–215: Photographs by and featuring Alexys Romo

217, 219, 221: Photographs by Hipolito Torres, featuring Taji Riley

218, 220: Photographs by Taji Riley

222, 224–227: Photographs by Zachary Kinsler, featuring Dominique Kinsler

230, 232: Photographs by Kyle Grossman

233: Photograph by Amber Grossman

235: Photograph by Kyle Grossman, featuring Amber Grossman

239: Photograph by Amber Grossman

240: Photograph by Kyle Grossman, featuring Amber Grossman

My vision of creating a gardening community wouldn't have been possible without each and every one of these women.
I would like to give thanks to all those who contributed their personal stories to this book and to all those who share their experiences and gardens with Black Girls Gardening online. It has been such a pleasure connecting with women from all over the world.

I would also like to thank my good friend Jillian, a certified herbalist, for helping me edit parts of the book. I couldn't have done this without you.

Thank you to my dear friend Maria for believing in me from the beginning and investing in BGG.

Thank you to my husband, Kyle, for allowing me to dig up parts of the yard and turn my gardening dreams into reality.

A special thanks to my brother, Dave.
My very best friend and the person I've always looked up to. All the time we spent together outside led me to where I am today.

My mother, Julie, my backbone, my support system. Thank you for all you do for me.
I love you endlessly.

Thank you to some of my close friends—Ceddy, Mal, Ally, Myranda, Catherine—and the rest of my family for supporting me and giving me words of encouragement along the way.

I will forever be grateful for the small town of Corning in upstate New York that drove me to create a space where I can always feel included and at home.

And to all my readers, I hope this book inspires you to connect back with the land, with your food, and with nature. Be patient and try not to compare your journey to others. Focus on what you want to accomplish in your space, and everything else will eventually fall into place!